CARL SAGAN

Recent Titles in Greenwood Biographies

Gloria Steinem: A Biography
Patricia Cronin Marcello

Billy Graham: A Biography
Roger Bruns

Emily Dickinson: A Biography
Connie Ann Kirk

Langston Hughes: A Biography
Laurie F. Leach

Fidel Castro: A Biography
Thomas M. Leonard

Oprah Winfrey: A Biography
Helen S. Garson

Mark Twain: A Biography
Connie Ann Kirk

Jack Kerouac: A Biography
Michael J. Dittman

Mother Teresa: A Biography
Meg Greene

Jane Addams: A Biography
Robin K. Berson

Rachel Carson: A Biography
Arlene R. Quaratiello

Desmond Tutu: A Biography
Steven D. Gish

Marie Curie: A Biography
Marilyn Bailey Ogilvie

Ralph Nader: A Biography
Patricia Cronin Marcello

CARL SAGAN

A Biography

Ray Spangenburg and Kit Moser

GREENWOOD BIOGRAPHIES

GREENWOOD PRESS
WESTPORT, CONNECTICUT · LONDON

Library of Congress Cataloging-in-Publication Data

Spangenburg, Ray, 1939–
Carl Sagan : a biography / Ray Spangenburg and Kit Moser.
p. cm.—(Greenwood biographies, ISSN 1540–4900)
Includes bibliographical references and index.
ISBN 0–313–32265–1 (alk. paper)
1. Sagan, Carl, 1934—Juvenile literature. 2. Astronomers—United States—Biography—Juvenile literature. I. Moser, Diane, 1944- II. Title. III. Series.
QB36.S15S63 2004
520'.92–dc22 2004015176

British Library Cataloguing in Publication Data is available.

Library of Congress Catalog Card Number: 2004015176
ISBN: 0–313–32265–1
ISSN: 1540–4900

First published in 2004

Greenwood Press, 88 Post Road West, Westport, CT 06881
An imprint of Greenwood Publishing Group, Inc.
www.greenwood.com

Printed in the United States of America

The paper used in this book complies with the Permanent Paper Standard issued by the National Information Standards Organization (Z39.48–1984).

10 9 8 7 6 5 4 3 2 1

To the memory of Isaac Asimov,
another candle in the darkness

CONTENTS

SERIES FOREWORD

In response to high school and public library needs, Greenwood developed this distinguished series of full-length biographies specifically for student use. Prepared by field experts and professionals, these engaging biographies are tailored for high school students who need challenging yet accessible biographies. Ideal for secondary school assignments, the length, format, and subject areas are designed to meet educators' requirements and students' interests.

Greenwood offers an extensive selection of biographies spanning all curriculum related subject areas including social studies, the sciences, literature and the arts, history and politics, as well as popular culture, covering public figures and famous personalities from all time periods and backgrounds, both historic and contemporary, who have made an impact on American and/or world culture. Greenwood biographies were chosen based on comprehensive feedback from librarians and educators. Consideration was given to both curriculum relevance and inherent interest. The result is an intriguing mix of the well known and the unexpected, the saints and sinners from long-ago history and contemporary pop culture. Readers will find a wide array of subject choices from fascinating crime figures like Al Capone to inspiring pioneers like Margaret Mead, from the greatest minds of our time like Stephen Hawking to the most amazing success stories of our day like J.K. Rowling.

While the emphasis is on fact, not glorification, the books are meant to be fun to read. Each volume provides in-depth information about the subject's life from birth through childhood, the teen years, and adulthood. A thorough account relates family background and education, traces personal and professional influences, and explores struggles, accomplishments, and contributions. A timeline highlights the most significant life events against a historical perspective. Bibliographies supplement the reference value of each volume.

ACKNOWLEDGMENTS

We could not have completed this book without a great deal of help for which we feel gratitude. A very special thank you to Ann Druyan for taking time out from her packed schedule to read the manuscript and discuss details with us. We also thank Pam Abbey of Carl Sagan Productions, Inc.; Alice S. Wessen of the JPL Office of Communications and Education; and Heather Lindsay of AIP Emilio Segrè Visual Archives. Also our appreciation goes to the following people at Greenwood Publishing Group: our editor Kevin Downing, who steadily encouraged us and saw the project to its end; editor Debra Adams, who started us out; and production editor Catherine Lyons. Thank you as well to Impressions Book and Journal Services. Also, to Jill Tarter and the late Bernard Oliver, we wish to express appreciation for past conversations that have contributed to our understanding of Sagan.

INTRODUCTION

Astronomer, planetary scientist, astrophysicist, exobiologist, educator, public figure, skeptic—all these hats represent important parts of Carl Sagan's complex, multifaceted career. Sagan offered to the world his extraordinary gift for cross-disciplinary research, his deep well of integrated visions and fruitful ideas, his vivid imagination, and his wealth of nonstop enthusiasm. He conveyed a sense of respect for the rigor and discipline of science to his listeners and readers. As his widow Ann Druyan puts it, he helped us all understand that "science is saying in the absence of evidence, we must withhold judgment" (Druyan, 2003, p. 29). Sagan arguably accomplished more for the public appreciation of science and scientists than any other person in our time. He conveyed to all of us the joys of asking questions and seeking answers, as well as the real and daunting challenges that confront the working scientist.

These accomplishments and his dedication to a demanding cause did not come without their price to both his family and his own personal life. Yet, in what could be written separately as a great love story, he found Druyan, with whom he fell deeply in love, and she with him. They were writing partners. They complemented each other in their strengths as writers, in their vision, in their personalities. Annie was caring and solicitous of his children. She brought out the gentler side of Sagan, softened what some observers considered his arrogance, and helped him to broaden his scope and mature. During their years together before his death, she helped him to become the Carl Sagan he set out to create, as a young boy standing on a hilltop looking up at the planets and the stars, and wondering…

TIMELINE OF EVENTS IN THE LIFE OF CARL SAGAN

9 November 1934	Carl Sagan is born in Brooklyn, New York,
1941	Sagan's sister, Carol Mae (Cari), is born.
1948	Sagan family moves to Rahway, New Jersey.
1951	Sagan enters the University of Chicago at age 16.
1954	Receives A.B. degree, with honors, University of Chicago.
1955	B.S. degree in physics, University of Chicago.
1956	M.S. in physics, University of Chicago; enters Ph.D. program in astronomy.
1957	Marries Lynn Alexander (who would later become Lynn Margulis, a distinguished biologist).
1957	The USSR launches Sputnik, the first artificial satellite.
1957	First scientific publication, "Radiation and the Origin of the Gene," appears in the journal *Evolution*.
1959	Birth of Lynn and Carl's son, Dorion Solomon Sagan.
1959	Spends the summer working with chemist Melvin Calvin at the University of California at Berkeley.
1959	In the fall, Sagan begins consulting for the National Academy of Sciences (NAS) and NASA on planetary missions, which he continues to do for the rest of his career.
1960	Birth of Lynn and Carl's son, Jeremy Ethan Sagan.

1960	Receives Ph.D. in astronomy and astrophysics from the University of Chicago; his thesis title: "Physical Studies of Planets."
1960	Fall, receives two-year Miller postdoctoral fellowship to research at the University of California at Berkeley.
Early 1960s	Publishes the theoretical model predicting an extremely hot surface on Venus; these conclusions are later shown to be accurate by a series of NASA missions to Venus.
1963	Joins the Harvard University faculty as assistant professor of astronomy; Lynn and Carl's divorce is finalized.
1960s	Furthers the research done by Stanley L. Miller and Harold Urey on the origins of life.
1968	Becomes director of the Laboratory for Planetary Studies at Cornell University in Ithaca, New York.
1968	Marries Linda Salzman.
1969–70	Linda and Carl collaborate on the design for a plaque carrying an interstellar message aboard *Pioneer 10* and *Pioneer 11*, spacecraft destined to exit the solar system.
1970	Receives promotion to full professor of astronomy and space science at Cornell University.
September 1970	Linda and Carl's son, Nicholas Julian Zapata Sagan, is born.
1970	Begins work on Viking, a mission to land two spacecraft on the surface of Mars; Sagan becomes the guiding light of the science team.
1973	First book, *The Cosmic Connection*, is published; it is a bestseller.
1976	Receives appointment as David Duncan Professor of Astronomy and Space Sciences at Cornell University.
1976	Sagan is asked to devise a symbolic interstellar message to be carried beyond the solar system by the *Voyager 1* and *Voyager 2* spacecraft.
1978	Sagan's book *The Dragons of Eden*, a discussion of the brain and human intelligence, receives the Pulitzer Prize for general nonfiction.

1980	Cowrites the *Cosmos* television series (13 episodes) with Ann Druyan; the public television series would be viewed by more than half-a-billion people in 60 countries.
1981	Linda and Carl's divorce is finalized; Carl marries Ann Druyan.
1982	Alexandra Rachel Druyan Sagan, the first child of Ann Druyan and Carl Sagan, is born.
1984	Birth of Sagan's grandson, Tonio Jerome Sagan, Dorion's son.
1985	Publication of novel *Contact*, coauthored with Ann Druyan.
1991	Birth of Samuel Democritus Druyan Sagan, Ann Druyan and Carl Sagan's son, Sagan's fifth and last child.
1992	Publication of *Shadows of Forgotten Ancestors*, second full collaboration with Ann Druyan.
1994	*Pale Blue Dot* is published.
1996	*The Demon-Haunted World: Science as a Candle in the Dark* is published.
20 December 1996	Carl Sagan dies.
1997	*Billions and Billions: Thoughts on Life and Death at the Brink of the Millennium*, Sagan's last book, a collection of essays, is published posthumously.
August 1997	The film *Contact*, starring Jodie Foster, opens.

Chapter 1

FROM BROOKLYN TO THE STARS: SCIENCE FICTION MEETS SCIENCE

Standing alone atop a high hill, the young boy looked up in wonder at the beckoning dots of light in the night sky. In the cool, clear air, the stars shone brightly as he sought out the planets. He knew where to find them, those mysterious other worlds the ancient Greeks had called "the wanderers." His gaze settled on one planet in particular, Mars—his special favorite, to him the most mysterious and intriguing of them all. For young Carl Sagan, no other planet held quite the same fascination as the one called Mars.

Maybe, just perhaps, if he stood on that hilltop long enough and wished hard enough, he would, like science-fiction hero John Carter in those otherworldly tales by Edgar Rice Burroughs, suddenly, effortlessly, find himself walking on the surface of Mars. Maybe he, too, would find strange Martian creatures, incredible gigantic plants, and exotic cities. But Carl knew that reality is more difficult than dreams, and that yearning alone would not whisk him away on a magical journey across the vast distance to the red planet. Wishes came true only in the fiction of pulp magazines and comics and between the covers of his well-thumbed Edgar Rice Burroughs books.

No, reality is much more demanding, and in reality the planets are far away, across the vastness of empty space; and wishes without action are no more than fruitless dreams.

Still though, Mars beckoned. And as the boy became a man he never gave up his quest or his belief that just possibly there might really be life on other worlds. Perhaps not the kinds of life pictured so colorfully by Burroughs—muscular heroic men and beautiful women—but some kind of life, exotic and unknown. For Carl Sagan the wonders of space and the mysteries of the planets would become a driving force in his life, a continual spark to his intellect, and a quest that would never be forgotten.

Perhaps the finest advocate for science of all time, Carl Sagan would also become one of the best-known scientists, and remain an individual with a vast and tireless vision. He would live to see human footprints on the Moon, and he would participate in landing robot spacecraft from Earth on the surface of Mars. He would contribute to the science behind the U.S. space program, fight for its use as a tool to further science, and thereby make his own mark as an esteemed scientist. And—perhaps his most priceless gift to humanity—he would provide the public with a front-row view of the beauty and the wonders of twentieth-century astronomy through the eyes of a scientist who loved his work. His was a tireless and highly successful campaign for the cause of science.

Born on November 9, 1934, Carl Edward Sagan was the first child of Samuel and Rachel Sagan. His father, Sam, immigrated to Brooklyn, New York, from Russia, having begun his life in the town of Kamenets Podolski in southwestern Ukraine. Sponsored by his 17-year-old uncle George, who had already made the journey, Sam left home for the United States when he was only five. His departure was well timed, because freedom for Jews in Ukraine was disappearing rapidly when he left in the early 1910s. The following decades would bring famine and the Holocaust murders would kill many thousands in Ukraine.

Sam's last name, Carl always claimed, derived from the founder of ancient Akkadian royalty, Sargon the Great, who conquered Mesopotamia (now Iraq), founded the Semitic Akkad dynasty, and ruled from about 2335–2279 B.C.E. Sam Sagan, though, never saw himself as a descendent of ancient royalty. But he did grow up with hopes for a professional career, and he enrolled at Columbia University to earn a degree in pharmacy. Known as "Red" (thanks to his bright-red hair) and "Lucky" (because during his college years he supported himself by beating fellow students at pool), Sam was an outgoing, affable man. The times worked against him, though. After the stock market crash in 1929, the Great Depression of the 1930s caused unemployment to soar. Thousands of destitute people stood in breadlines to get free meals. Sam found himself forced to drop out of school, give up his career dream, and get a job instead of getting ahead. He was lucky, though. He found a job as an usher at a movie theater—though it didn't pay much—and later went to work for the New York Girls' and Women's Coat Company as a garment cutter.

Samuel met the love of his life, Rachel Gruber, at a party. Sophisticated, smart, and vivacious, she enchanted him immediately. On March

4, 1933, the two married in Brooklyn. As their daughter Cari Sagan Greene (Carl's younger sister) remarked many years later, "Mother and Father were always very close. They were very much in love" (*A&E Biography*, 2000). They were also an illustration of the adage that opposites attract.

Rachel had a storied background. Her father, Leib Gruber, fled from what is now Ukraine to New York in 1904, with the law at his heels. The charge was serious—he had killed a man over an issue of honor, and he had to leave fast, even though he had no money to pay for his wife's passage. So he had to sail without her. He was able to send for her the following year, though, and Chaya Gruber arrived at Ellis Island with a dollar in her pocket and a heart condition. They made their home on East Seventh Street in Manhattan's Lower East Side, where many other Jewish immigrant families had settled and the tenement housing was affordable, if unpleasant. Chaya was eager to become American, changing her name to Clara and naming her firstborn Rachel Molly Gruber. But Rachel never really knew her mother; about two years later, Clara Gruber died shortly after the birth of Rachel's younger sister. Heartbroken and unable to raise two daughters by himself, Leib sent Rachel to live with her aunts in Austria. There, Rachel would almost certainly have suffered in the Holocaust had Leib Gruber not arranged for her return when she was four, shortly after he remarried. Rachel returned to the United States to a safer but emotionally troublesome life—always tinged with the sense of being second best, after her stepsiblings, in the eyes of her stepmother.

By the time Rachel had grown up and married Sam, she had become a dazzling, sometimes impatient, and emotionally turbulent young woman—always a bit mysterious because of her complexity. She was also eager to get ahead, competitive, intelligent, well read, and known for her quick, incisive wit.

Sam, on the other hand, was laid back, self-effacing, and always eager to please her. He liked to kid around, and he was supportive to his children and interested in their mental and emotional growth. His ambitions to become a pharmacist shelved forever, he would move up the ladder into management at the coat company and make a comfortable living for his family. Rachel never wanted for coats, for which she characteristically had matching gloves, hats, and shoes.

When Carl was born, his parents probably chose his name to honor his grandmother—using an Americanized, male version of Clara or Chaya. Or, his name may be a shortened rendition of his mother's grandfather's name, Kalonymous. Carl's middle name, Edward, had a romantic origin—

honoring the former Edward VIII, king of England (later the Duke of Windsor), who gave up the throne of England out of love for Wallis Warfield Simpson, a divorced American woman whom the British Parliament would not accept as a member of the royal family. Perhaps the name was Sam and Rachel's homage to their own love.

For Carl, beginning life in Brooklyn in 1934 may not have been the best of times and places, but it was not the worst. His father had work and business was decent, an advantage not everyone shared in the mid-1930s, smack in the middle of the Great Depression. The Sagan family lived in a series of unpretentious apartments in the Bensonhurst area of Brooklyn, a tidy neighborhood populated primarily by residents of Jewish or Italian descent. Local truck farmers used scattered open spaces to grow their crops. Sam Sagan didn't bring home a lot of money, but Rachel was both frugal and adept at stretching the household budget. She took her role as mother and housewife seriously and strove to serve nutritious, balanced meals, based on the free U.S. Department of Agriculture booklet on the subject. Also serious about her cultural heritage and religion, she cooked according to Judaism's strict Kosher regulations and her menus were homey, balanced, and nutritious. If she served fish for dinner, she would also have vegetables—perhaps spinach or other greens smothered in butter—with pudding for dessert.

People who knew Rachel seem in complete agreement on one central point: She doted on her firstborn from the moment she first held him in her arms. For the rest of Rachel's life, Carl was the center of her world. As one friend of Carl's humorously put it later, "She worshipped the ground he floated above" (Davidson, 1999, p. 9). She avidly researched questions on children's health in government booklets on the subject, underlining important points. She concerned herself with the quality of Carl's education. And she was always ambitious for him.

At the same time, Rachel was often sharp-edged and insecure. Her own early life had taught her the world was an uncertain place at best—run by unpredictable fate, the same fate that had snatched Rachel's mother away and packed her as a young child off to Europe to live with strangers. The same fate that had saddled her with a life of competition against her step-siblings. She took on the world as a giant challenge, "her expression, implacable," a face that presented "her dare to the world to get in her way" (Druyan, 2000).

In later years, Carl liked to recall two incidents from his childhood—events he defined as examples of key lessons he learned from his parents. "My parents were not scientists," he wrote. "They knew almost nothing about science. But in introducing me simultaneously to skepticism and to wonder, they taught me the two uneasily cohabiting modes of thought that are central to the scientific method" (1996, p. xiii).

He learned the first from his mother one wintry evening in 1939 as they waited for Sam Sagan to return home for dinner. Rachel had changed her clothes and put on makeup. Together, mother and son looked out the apartment window as the sun set on the wintry waters of lower New York Bay.

"There are people fighting out there, killing each other," she remarked somberly.

Relatives and friends in Europe were already in grave danger. German Nazi troops had invaded Poland and the chilling fear of violent anti-Semitism loomed throughout Central Europe. On November 9, 1938, Adolf Hitler's propaganda minister, Joseph Goebbels, had orchestrated the burning of nearly two hundred Jewish synagogues and the vandalism of Jewish shops and businesses. By September 1939, World War II had begun.

"I know," said five-year-old Carl, just as seriously. "I can see them."

"No, you can't," she retorted, almost an edge in her voice. "They're too far away." Maybe she was annoyed at the intrusion of the boy's naïve imagination into the stark reality. Maybe she wanted to protect him from the implication—even in the world of his imagination—that the danger could be so near.

In Sagan's account of this moment given in *The Demon-Haunted World* (1996, p. xii), Sagan reflects on his childhood dismay at the near-impatience he heard in the words of his mother, who generally expressed only love and approval for her young son. He thought he had seen figures far away on a distant shore. How could she be so sure that he couldn't? It was an early lesson to the future scientist that he never forgot: No one is easier to fool than oneself—because sometimes we see only what we want to see, not what is really there. The soldiers in Europe were much too far away to be seen—across an ocean of water. Imagination and reality, he was discovering, are often in conflict and what you think you see is not always real—lessons in skepticism that may have been difficult for such a young child.

Sam was the one who taught Carl the wonder of numbers—a characteristic so identified with Sagan that decades later cartoonist Gary Larson would caption a cartoon, "Carl Sagan as a kid." The boy is standing with a girl on a hilltop looking up at the stars, pointing, and the dialogue balloon says, "Just look at all those stars, Becky... There must be *hundreds* of 'em." Sam showed Carl how zero acts as a placeholder when making calculations, and he talked about large numbers. Carl decided, with a five-year-old's enthusiasm, that he would write all the numbers between 1 and 1,000. Sam knew this was a good project, and for Carl's carefully drawn pencil marks he supplied pieces of gray cardboard (probably from the coat factory). Bath time and Rachel arrived, and Carl was nowhere close to finishing. Sam saw from Rachel's face that the bath could not be put off. So he devised a compromise. Why not let him take over writing numbers while Carl took a bath? Then the boy could finish up the job before bedtime. By the time Carl was back, his father had closed the gap from the low hundreds to nearly nine hundred. Carl easily finished the job and went to bed, no doubt still marveling at the beauty of really large numbers. It was more than a job finished: It was another beginning.

THE EDGE OF TOMORROW

Despite the continuing economic depression and the war, 1939 was a time of great optimism in the United States. Technology had transformed people's lives since the turn of the century. Five-year-old Carl's grandparents had grown up, not only on the other side of the world, but also in a time when neither telephones nor electricity served people's homes—in either Ukraine or the United States. In the world of his grandparents' youth, no one had ever flown an airplane. Automobiles were only just being invented, and communication by letter was slow—delivered by train or transoceanic ships. Telegraphy was a struggling invention, no one owned radios, "moving pictures" were novelties, and the synchronized audio of "talkies" had not yet arrived. But things were vastly different by the time Carl was born, and by 1939 airmail and air travel were already common. Radio programs formed an important part of life in Greater New York City. Kids in Carl's generation would also make weekly trips to the movies to take in Saturday matinees. (Television was not a widespread presence in living rooms until the 1950s.) Electricity ran refrigerators, lamps, stoves, and washing machines in people's homes, and telephones connected families across thousands of miles.

No greater symbol of this sense of progress marked the era than the 1939 New York World's Fair. This vibrant international exhibition was an extravaganza—a vividly enthusiastic look at the exploits of progress yet to come. For the Sagan family, it was practically in their backyard, and they were among the millions of spectators who thronged to see what tomorrow would look like. As an adult, Sagan recalled attending the fair with his parents, riding on his father's shoulders, and taking in the wonders. This early exposure to the promise and excitement of technology and science captured his impressionable imagination.

From infancy, Carl showed the curiosity and thirst to know and discover that infused his entire life and lit the candles of his sense of wonder. He was filled with questions, endless questions, about everything. "What are the stars?" he asked. But no one he knew had the answer. Had he wanted to know about the machine his father used to cut garment pieces out of cloth, Sam Sagan would gladly have answered. Had his question centered on a topic his mother had covered in her own thirsty reading, she would joyfully have explained. But no one Carl knew had a good answer to this question, so his mother set out to solve that problem. She took him to the Brooklyn Public Library and introduced him to the librarian, then waited patiently while Carl struggled with a facial tic that often got in the way of his words as a boy. Finally he asked for a book about the stars. The librarian knew what he wanted. She disappeared for a few moments, then returned with a book and handed it to Carl. He looked at it, puzzled. It was about Hollywood actors, far from what he expected. So Carl explained carefully what he really meant: the stars in the sky—not the ones in the movies.

That day was the beginning of a long voyage—an intellectual voyage that required the same determination he showed that day in the library. It was a voyage to the farthest reaches of the universe, a love affair with science and its methods. It was the ultimate adventure, to him, the greatest human quest, to *find out*—surely and without fooling oneself—what the universe is really like.

REALITY VS. IMAGINATION

The kids in Carl's neighborhood remembered him later as standoffish and reserved. But there was one kid for whom he would always seem a hero. Carl turned seven in 1941, and that same year another member of the family arrived. With their usual respect for Carl, Rachel and Sam allowed him to name his new little sister, and, apparently never threatened with thoughts of sibling rivalry, Carl sought to come up with a name that would be as close as possible to his own. The name he settled on was Carol, and so she received the name Carol Mae Sagan. Later she became known as Cari.

One day when Cari may have been about five, she was out riding her tricycle. A friend dared her to ride down a steep street, and her trike went out of control. Cari fell and was badly hurt. The next thing she knew, Carl was standing over her. He put his arms around her, picked her up gently, and carried her home. In Cari's mind, Carl was her wonderful big brother, her protector.

Carl played the standard kids' games of the time with his friends—cops and robbers, cowboys and Indians, or American soldiers vs. the Nazis. Coney Island Beach was nearby, and the kids, including Cari, went there together. Carl's friend Robert Gritz tells of ogling French postcards with Carl or listening to classical music, such as the *William Tell Overture* (with *The Lone Ranger* theme that every kid loved). Kids' radio programs that aired on Saturday mornings and early evenings enticed Carl and Robert to sit listening to adventure programs starring powerful enemies of injustice and crime, such as *Captain Midnight* and *Superman*, a special favorite of Carl's and a precursor to the later TV series and movies. One program debunked mysteries and occult enigmas by showing natural, logical explanations. This influence may have laid the groundwork for the strong stands Sagan later took against superstition, the paranormal, and "demons," as he referred to all irrational, nonnaturalistic beliefs.

Carl and Robert also experimented at a young age (six or seven) with lenses—discovering that two held together enabled them to examine the Moon and its craters and the color of Mars. They also went to the Museum of Natural History in New York City, where they visited the Hayden Planetarium. Carl was feeding his growing love for astronomy.

Sagan later recalled that he was held spellbound by the dioramas on his visits to the museum (1996, p. 347). These fascinating displays transfixed observers by setting figures of animals in the natural scenery of their habitats. Visitors walked from one brightly lit scene to the next in the darkened hall, experiencing now a peaceful tableau of mule deer nibbling vegetation in North America, and in a few steps crossing continents to view, say, Bengal tigers nursing their young. The same museum featured exhibits of dinosaurs, featuring huge reconstructions of dinosaur skeletons—and these set Carl to devouring everything he could find on that subject as well.

New York City, with its large Jewish population, was hard hit emotionally by World War II and the horrors of the Holocaust. But Rachel—who took the news from Europe very hard—protected Carl from the grim facts of slaughter, even of the Sagans' relatives. One biographer hypothesizes this very protection enabled him to become an eternal optimist (Davidson, 1999, p. 15). Others might argue that his optimism was innate—part of a genetic legacy he received from forbears who came to America filled with hope. In either case, Sagan's positive outlook both provided him with emotional strength and infused the message he communicated to the world. But it would also blind him at times to the realities of politics and human nature.

That blinding optimism extended to the greatest of political fears following World War II: the specter of nuclear attack. With a few powerful nuclear blasts, human beings could now destroy the world. But young Carl, who was 10 years old at the time, looked upon atomic power with more hope: Perhaps this is the way to reach Mars, he thought—using rockets powered by atoms. Not until much later in life did he finally give full due to the deadly Pandora's box opened by the dropping of atomic bombs on Hiroshima and Nagasaki in Japan during that last summer of the war.

The imaginative world of science fiction, however, with its vision of space exploration, fueled Carl's future as a scientist. Few scientists or inventors born in the first half of the twentieth century fail to credit science fiction with a large measure of influence on their choice of careers. These tales of imaginative adventure threw off many of the restraints of reality and explored possibilities, real or unreal, that might exist in the future, in some other universe, or some other planet. Carl Sagan was no exception. By the time he was eight, he had already begun to consume the science-fiction works of Edgar Rice Burroughs. These books sported vividly colored portraits of beings from other planets and Earthly heroes in combat against unspeakable monsters. Titles such as *A Princess of Mars*, *The Gods of Mars*, and *The Warlord of Mars* spoke of extraterrestrial royalty, deities, and power. These books featured sumptuous women and the hero John Carter, a strong, muscular man of vast, hidden powers. Most intriguing of all, the inhabitants of Mars came in a range of types—including four-armed humanoids with scales and antlike heads. Of course, intelligent beings who lived there called the place by their own name, "Barsoom."

Were there really living beings on Mars? Was the great astronomer Percival Lowell right after all? Lowell had thought he saw canals on Mars through his telescope—vast networks of constructed water troughs crossing the planet's surface. He published his hypothesis in 1906. No one else really saw them, though, and other scientists later demonstrated the canals had been optical illusions. Yet, Lowell's Martian canals had captured the imagination of other scientists and the public alike, and Burroughs's books built on that enthusiasm. As a kid, Carl caught Mars fever early and he never ceased to be fascinated by the red planet, whose reddish surface reflects a steady light from an average distance of about 48 million miles from Earth.

Today no one—except perhaps for the most uncontrollably imaginative—thinks there are canals on Mars, but many scientists entertained the possibility up until robot spacecraft orbited the planet and mapped its surface. If the canals had been there, they would have shown up, and when they didn't, their existence was finally, conclusively ruled out. But not the possibility of life on Mars. Today, many scientists think some remnant fossil of primeval microbial life may remain on the red planet from a time when the beginnings of life may well have existed there. The question of life on Mars continued to tantalize Sagan throughout his life, and a solid cadre of his fellow scientists continues the search.

CREATING CARL SAGAN

Sagan's mother had a vision for her son that was filled with ambition and possibility, and her expectations unquestionably influenced him. However, even as a boy he slowly developed a view of himself that was unique. He was a cerebral, imaginative boy who developed a love for thinking and writing and optimism. A friend of Sagan's once said, looking back over the years she had known him, "He created the person he wanted to be" (*A&E Biography*, 2000). And he set out early on that path.

When Carl was about to begin high school, Samuel Sagan was promoted to manager of a new plant of the New York Girls' and Women's Coat Company, opening in Perth Amboy, New Jersey. So in 1948 the Sagan family pulled up its roots and moved to Rahway, New Jersey. Their new house must have seemed luxurious after the relatively cramped quarters in their Brooklyn apartment. It was a two-story clapboard structure with a front porch and a yard. It stood in a middle-class town with the charming historical claim that George Washington had breakfasted on ham and eggs at the Merchants and Drovers Tavern (which is still there today). They lived at 576 Bryant Street. To the Sagans their new home must have seemed like a palace.

Rahway High School, though, was a disappointment. Carl's teachers did not inspire him and he was bored. However, he didn't fail to impress those same teachers he considered so mediocre. His continuing fascination with the stars and the constellations led him to soak up all he could on the Greek and Roman myths relating to the connect-the-dot shapes imagined by early sky watchers. After filling classroom blackboards one day with notes and illustrations during an exhaustive class presentation he made on the myths and lore of classical Greece and Rome, his teachers called his parents in to visit the school. This boy is gifted, they told Samuel and Rachel. He needs the stimulation of a school for the gifted. But for some reason, possibly financial, the Sagans kept Carl at Rahway High, where he completed his secondary education.

However, Carl's curiosity continued to prod him to soak up everything he could about science in general and astronomy in particular. He sent for a book advertised in the pulp magazine *Astounding Science Fiction*. He waited impatiently for the book to arrive. Like the adventuresome fiction of Edgar Rice Burroughs, the book, *Interplanetary Flight,* struck a chord in young Sagan. But this book was not an adventure story of make-believe feats. Written by the respected British astronomer Arthur C. Clarke, *Interplanetary Flight* spoke of real possibilities (even though many people did not understand that at the time). Clarke was serious. In Carl, they stirred a real and palpable excitement. He began to see a future in which rocket power might make such a possibility a reality.

Now for his fiction reading Carl began to consume works written by other scientists who also wrote science fiction—J.B.S. Haldane, Julian Huxley, Arthur Eddington, Sir James Jeans. These stories were not based on fantasy; they were based on feats that would one day be possible. Since 1914, American rocketeer Robert Goddard had been building and firing rockets and tested the first successful liquid-fuel rocket in 1926. In Germany, the V-2 ballistic missile developed for the German military by Wernher von Braun and his team soared to the edge of the atmosphere before plunging to Earth to destroy sections of London, other regions of England, and the Netherlands during World War II. As Carl was completing high school, these engineers shifted their allegiance and citizenship to the United States and were improving the range of their rockets. By 1957, scientists from the Soviet Union would place the first satellite, *Sputnik I,* in orbit—and thereby jump-start the space age. The Soviets would follow soon after with a second satellite. By January 1958, von Braun and his team would launch the first U.S. satellite. These two events had been brewing for a long time—fed for centuries by wishful thinking and in the previous half century by verifiable calculations. Human-made machines were venturing beyond Earth's atmosphere for the first time.

Young Carl was one of those who started out very close to the ground floor of this singular chapter in human existence. He began to soak up nonfiction works by 1940s scientists who wrote for the public—including Rachel Carson (on biology), George Gamow (on rockets), Willy Ley (on early rocket history), and Simon Newcomb (on astronomy). He began to see that if he wanted to be a part of the coming era, he would need to learn much more about science and even—he reluctantly recognized—mathematics in general and calculus in particular.

UNIVERSITY YEARS

Sagan may have been bored in high school, but his university choice, the University of Chicago, proved to offer exactly the kind of broad-based, thoroughgoing education that could meet the needs of the eclectic visionary he would become. At age 16 in the fall of 1951, Sagan entered the freshman class. It was a time when the undergraduate curriculum at Chicago required an across-the-board liberal arts education without focusing on any particular academic field until graduate school. In an innovative curriculum designed by university chancellor Robert Hutchins, students read the great books of knowledge, tracing the growth of ideas by reading original sources. So, though Sagan would eventually major in physics, he developed a broad knowledge of literature, art, history, music, and other diverse disciplines. It was the perfect intellectual cross-training for the mental athletics required of the future interdisciplinary star.

Sagan's extracurricular activities also developed tools for the future, including active participation in the astronomy club. Always interested in theory more than other aspects of science (such as experimentation and observation), Sagan became head of the club's theoretical branch. In that capacity, he arranged for physicists and astronomers to talk to the group, and thereby had the opportunity to speak one-on-one, if briefly, with such major theoreticians as Subrahmanyan Chandrasekhar (now also the namesake of NASA's X-ray space telescope, *Chandra*).

INTENSITY AND OPPORTUNITY

For many students, university or college life is the first experience away from home—both an exhilarating and stressful period. Additionally, the leap from high school to university classes is usually not an easy one. Demands by instructors are often high, and serious students see their university years as a direct preparation and qualification for a lifetime career. Carl Sagan's university years took a toll. During his first years at Chicago, Sagan developed a disorder known as achalasia (which literally means failure to relax), or cardiospasm, a condition that interferes with the normal muscle processes that move food through the tube structure of the esophagus to the stomach. It is caused when the lower esophageal sphincter fails to relax (and so the name achalasia). Resulting problems usually include difficulty in swallowing and even breathing, as well as chest pain and regurgitation of food—alarming symptoms that could cause choking and even death. According to his sister Cari, the condition may also have had "something to do with—Rachel. And [his] intenseness" (Davidson, 1999, p. 42). An intense young man, Sagan may have internalized many of the stresses inherent in college studies and the effort required for success. On the other hand, the condition can be brought on by other conditions or heredity. At 18, Sagan traveled on his own to the Mayo Clinic in Minnesota to investigate ways to cure or control his condition. There, physicians attempted to stretch his esophagus, but the procedure did not solve Carl's problem and he continued to have trouble for many years to the point that as a young man he dreaded eating, for fear he would not be able to swallow—probably contributing to his thin, gawky appearance in his early adult years.

Perhaps even more important to Sagan than eating, though, was learning, and a golden opportunity came to him through an insightful suggestion made by his mother when he was home for a visit during his first year in college. She had discovered the nephew of a friend was also in town visiting his uncle, and it so happened the young man was a graduate student in biology at the University of Indiana in Bloomington. She had a hunch her son might hit it off with him and at the very least obtain some insights from a peer with similar career goals, so she urged Carl to pay him a visit.

At the time, Carl was more intent on a game of one-on-one basketball, but his mother pressed and he went. The occasion validated Rachel Sagan's insight. Carl found much to talk about with Seymour Abrahamson as the two quickly discovered they shared an interest in how life began. This topic was closely connected to another question that intrigued them both: Does life exist only on Earth, or is it easily formed in the universe? What if molecules readily formed into the basic ingredients for life on primeval Earth? Then wouldn't one logically conclude life very likely also formed on other planets revolving around other stars in our Galaxy and beyond? Would not the same or a similar set of natural circumstances occur elsewhere? These were tall questions to ask for the time—in the days prior to the dawn of the space age. No one had yet launched a satellite, much less traveled in space. Nor had anyone yet discovered a planetary system beyond our own.

As it turned out, Abrahamson was studying under Hermann Joseph Muller, the recipient of the 1946 Nobel Prize for physiology or medicine honoring his work on the genetics of the fruit fly (*Drosophila*). And Muller was also interested in the origins of life, as a sideline. Abrahamson apparently shared some of Sagan's ideas on the subject with Muller, and during a visit to Bloomington, Sagan and Muller met. The enthusiastic student and the Nobelist immediately saw ways they were in sync.

Muller was open to the idea of life in other galaxies—of Darwinian evolution beyond Earth's tiny cradle, a position rejected outright by most scientists at the time. He not only generated excitement about science, its purpose, and its methods, but he also encouraged Sagan and was willing to listen and discuss possibilities. The visit culminated in another exciting opportunity: Would Sagan like to work for Muller during the summer months between academic years? Sagan jumped at the chance.

So Abrahamson's friendship led Sagan to an avenue of science physicists rarely explore. This experience added to his eclectic knowledge and served as the foundation for his understanding of the scientific investigation into biological fundamentals. For a student who would eventually pursue the question of how life began, this moment was key. That summer in Indiana Sagan worked with his first mentor, a world-class scientific mind. He also delved into his first work as a scientist.

Born in New York City on December 21, 1890, Hermann Joseph Muller was the paternal grandson of an artisan who left his native Rhineland in 1848 along with his family to seek greater freedom in the United States. His mother's ancestors, who were Spanish and Portuguese Jews, had fled the Inquisition and traveled north to Britain and Ireland generations earlier. Her father and mother emigrated from Britain to New York City, where they settled. Muller's father died the year Muller turned 10, but he passed his curiosity about nature, evolution, and the universe, as well as his concern for human welfare, on to his son—interests also encouraged by his mother.

Muller attended public school in Harlem and high school in the Bronx, organizing, with his friends, what was probably the first high school science club. Young Muller received a scholarship, based on his examination marks, to Columbia University. But to stay ahead he worked long hours for low pay as a bank runner and hotel clerk.

In college he was quickly drawn to biology and then specifically to genetics. He performed graduate work at Cornell Medical College, followed by several years as a teaching assistant at Columbia. These were the days when Thomas Hunt Morgan's graduate students were engaged in groundbreaking work in genetics via the fruit fly (*Drosophila*). This humble insect became the experimental creature of choice for geneticists because of its extremely short life span. Multiple generations can be studied within a relatively short time, making the legacy of an individual's genetics more quickly apparent. In what came to be called the "Fly Room," Morgan and his students were able to mate flies with specific characteristics and to observe how they were passed on to offspring. Muller formed part of this group. However, he did not begin his own experimental work until 1912.

Muller pursued his research while serving on a series of different faculties beginning with Columbia, then Rice Institute in Houston, Texas, then Columbia again, and then the University of Texas at Austin. His work on several genetics projects culminated in 1926 when he demonstrated that X-rays can cause abundant gene mutations and structural changes in chromosomes. He published his results in 1927, but he didn't receive the Nobel Prize for his work until 1946—about six years before Sagan met him.

Muller continued to research and teach genetics on faculties in several countries, including the Institute of Genetics of the Academy of Sciences of the Union of Soviet Socialist Republics (USSR) in Leningrad, and later Moscow. But as the geneticist Trofim Denisovich Lysenko, who proposed that acquired characteristics could be inherited, gained a lock on Soviet policy (with some disastrous results when applied to farming), Muller returned to the United States to teach at Amherst College in Massachusetts from 1940 to 1945. In 1946 he joined the faculty at the University of Indiana at Bloomington, where he and Sagan crossed paths a few years later.

During Sagan's summer at Indiana, he cemented his friendship with Muller. At the same time he also had the opportunity to find out once-and-for-all that the experimental laboratory was not his cup of tea. He found the work he was asked to do boring and tedious. In Muller's laboratory, rigor was required, attention to detail, expected. And the details Sagan was expected to observe were tiny and easily overlooked.

Muller's contribution to Sagan's early confidence, though, was large. Muller held a broad education, was supportive, and, like Sagan, enjoyed science fiction. He was also not stuffy outside his laboratory—he even attended a science fiction convention with Sagan. And he listened to Sagan's exploration of ideas.

Sagan had supreme confidence, even as an undergraduate. He just walked up to people, regardless of their reputations, and deftly engaged them in conversation. Strikingly articulate, he spoke with authority and ease. He had interesting ideas and asked good questions. His eloquence had the rare quality of substance—even though in those early days his ideas were still ripening. He was, after all, still a very young—though serious-minded—individual.

From this fearless willingness to put himself on the line evolved a group of priceless friendships—relationships that turned into further opportunities for development as a scientist and as a thinker. In addition to the summer he spent with Muller in 1952, he used the summer months in 1956 to work with planetary scientist Gerard Kuiper (at the time the *only* full-time planetary astronomer in the world), followed by a summer in 1957 with Nobel Prize–winning physicist George Gamow in Colorado, and in 1959 in California with Melvin Calvin, who would win the Nobel Prize in chemistry two years later for his work with photosynthesis. In a much less structured way, the mentor who was most important to Sagan's acceptance in key circles was a young geneticist named Joshua Lederberg, who, at 33, would share the Nobel Prize for physiology or medicine in 1958 for his work in genetic structure and function in microorganisms.

Not many students are lucky enough to attract the attention of such important mentors. But with his unflagging intellectual curiosity, fearsome eloquence, and enthusiastic charm (which would later captivate millions), Sagan created his own luck.

Chapter 2

HOW—AND WHERE—DID LIFE BEGIN?

As a boy, Sagan was enchanted by the adventuresome novels of Edgar Rice Burroughs, set on the exotic surface of "Barsoom." But Burroughs's novels also raised serious and fundamental questions in Sagan's mind—and established Sagan's lifelong yearning to know about other worlds, the nature of life, how life began on Earth, and whether life existed anywhere else in the universe. These questions would form the basis of a new discipline, not yet formed when Sagan was pursuing his education. This discipline would become known as "exobiology," and later as "astrobiology," that is, the study of biology and life as a universal possibility on this planet and on other worlds. These questions intrigued Sagan throughout his life. What were the real possibilities? How would life begin on another planet? How, for that matter, did life begin on Earth, in terms of chemistry and physics? In the beginning on Earth, conditions would have been most inhospitable—so what happened? How, exactly, did a lifeless planet—prebiotic Earth—become a wonderland of diversely elegant life-forms? His imagination awakened by Burroughs and other science-fiction writers, Carl Sagan found himself exploring real, verifiable, scientific answers to these questions throughout his life. This melding between a passionate appreciation of the romanticism and glory of the universe and a respect for the stern discipline of science would become one of Sagan's major strengths as science advocate.

ORIGINS OF LIFE: A TIMELESS QUESTION

Sagan soon discovered he was not the only one who found these questions intriguing and exciting. Even though the terms "exobiology" and "astrobiology" did not even exist when Carl Sagan began taking classes at the University of Chicago, several chemists, biologists, and astronomers were beginning to very seriously consider questions regarding the origins of life. Sagan was in exactly the right place at the right time to see their work firsthand. Soaking up a headily eclectic series of subjects, Sagan explored fields most physics majors would find completely alien. Perhaps Sagan could thank Chicago's broad, humanities-oriented undergraduate curriculum for this. Perhaps he could thank his own broad interests. Either way, over the next few years, he rubbed elbows with some of the greatest thinkers in several different fields and each one contributed to his growth as a scientist.

Pondering the origins of life was certainly not a new human pursuit, however. For millennia people have wondered about how life began on Earth. For the most part, their explanations have been mythological or religious, using symbols and parables that added color, texture, and human meaning to the answers they supplied. The most sophisticated explanations have incorporated all-encompassing worldviews that seek to integrate experience and interpretation. But throughout history, the essence of life has eluded our understanding. Where did life come from? How did life begin on Earth? What essential difference allows a plant to grow and replicate, but not a stone? Biological life has always been a mystery to humans—considered special and beyond human understanding.

Slowly, another thread began to be woven into the tapestry of knowledge, however. A more secular view, based on observation and experiment, began to develop. Using the tools of science, humans began to discover new evidence regarding how Earth came to be and how it came to be inhabited by living organisms that reproduce, or copy, themselves (replication) and make use of energy from the outside world (metabolism) to engage in such activities as growing, reproducing, moving, and responding to the environment.

This process of discovery has taken centuries and is by no means over. New insights and discoveries are continually coming to light. But they all fit into a scientific set of explanations that does not require a special force to start things going. Humans, animals, plants, all things living and nonliving are, as Sagan liked to say, "star stuff." We and our neighbors on this planet are made of the same elements, the same chemicals as the stars and other matter in the universe.

This view of living processes and ingredients as ordinary (though fascinating) chemistry and physics began to form as early as the eighteenth century, when Luigi Galvani (1737–98) touched the leg of a dead frog with an electric charge and found the leg reacted by twitching, even without the presence of life. He realized that some process not restricted to living things must be involved.

In the following century, Friedrich Wöhler (1800–82) succeeded in synthesizing the organic compound *urea*, which is abundant in the urine of humans and mammals. At the time, organic compounds were thought to be impossible to create in a lab. It was assumed that anything created by an organic process was essentially different, "vital," and could not be created with inorganic materials. Wöhler's synthesis of urea put an end to the idea that organic compounds differed fundamentally from inorganic compounds.

So by the time Sagan began his studies at Chicago, several centuries of scientists had already traveled far in the journey toward understanding the physical laws operating in the universe, the genetic processes governing replication, and more. But even from our current vantage point at the start of the twenty-first century, we cannot answer two big questions: Did life begin on its own, or did it have a mystical origin? And does life of any kind exist on other worlds? These major, fundamental questions remain unanswered for the moment, even though Sagan and his colleagues believed the answers would be around the next corner.

In the fourth century B.C.E., Aristotle argued that living things spring spontaneously out of inorganic substances, and for centuries afterward, scientists tried to prove him right or wrong. In the nineteenth century, it seemed, Louis Pasteur had finally provided a definitive answer: Spontaneous generation could not take place. His trick was to use a special, swan-necked flask that kept out contaminants such as plant or mold spores, while allowing normal atmospheric conditions. By correcting flaws in previous experiments that had seemed to prove otherwise, Pasteur demonstrated that a completely sterile, inorganic solution would not produce any signs of life, even in the presence of all the requirements of life (such as hospitable temperature, presence of oxygen, and so on). Or so it seemed.

But a young graduate student named Stanley Lloyd Miller made an important breakthrough in the early 1950s at the University of Chicago, right under Sagan's nose. Miller was working on his Ph.D. under the guidance of Harold Clayton Urey (1893–1981), a noted American chemist who received the Nobel Prize for chemistry in 1934 for discovering heavy hydrogen (deuterium).

H.C. Urey's grandparents were early settlers in Indiana, and he was born in Walkerton on April 29, 1893. He attended local country schools, becoming a rural schoolteacher himself in 1911. Three years later he entered the University of Montana, earning a bachelor of science degree in zoology in 1917. He then worked as an industrial research chemist while teaching at the University of Montana before returning to graduate school at the University of California in 1921 and receiving his Ph.D. in chemistry in 1923. A year of postgraduate work at the Niels Bohr Institute in Copenhagen followed. On his return to the United States, he joined the faculty at Johns Hopkins University, also serving at Columbia University from 1940–45 as director of war research leading up to and during the Manhattan Project, which produced the first atomic bomb.

Urey's early research revolved around the entropy of diatomic gases and the problems of atomic structure, absorption spectra, and the structure of molecules. However, the research for which he is best remembered is an experiment he oversaw in the early 1950s, during the time Sagan was attending the University of Chicago.

Urey had become interested in geochemistry, the formation of planets, and atmospheric conditions during the beginning of Earth's history. He also began to wonder about Pasteur's definitive proof that spontaneous generation could not occur under any circumstances. What if, Urey and Miller asked, instead of waiting just four days, Pasteur had waited the billion years or so that Earth had waited for the first life to appear? And what if, instead of the current mix of nitrogen and oxygen, he had tried to simulate the primordial atmosphere that existed in the first few billion years of our planet's existence? What if he had used an ocean full of inorganic molecules?

With Earth's age fairly reliably dated at 4.55 billion years, and the first evidence of life dating back to about 3.5 billion years in Western Australia (or possibly earlier, according to recent discoveries), where and how, exactly, did prebiotic Earth, containing no life, become biotic Earth, teeming with life?

Urey had begun to think an experiment could be interesting. The idea was to begin answering the age-old question, "How did life begin?" Could life have begun spontaneously? Urey was reasonably sure that Earth's primordial atmosphere was quite different from today's atmosphere. He estimated it was most likely made up almost entirely of hydrogen-containing gases such as methane (CH_4), ammonia (NH_3), and water vapor (H_2O). The proposition that hydrogen might have dominated the early Earth atmosphere was an insight that would not have occurred to anyone 50 years earlier. For this, Urey was indebted to the insights of an astrophysicist named Cecilia Payne-Gaposchkin, who taught at Harvard in the 1920s.

In 1926, British physicist Sir Arthur Eddington was trying to work out the mystery of what powered the Sun's enormous energy, and basing his work on Einstein's famous equation, $e = mc^2$, he put the first piece of the puzzle in place. Since *e* stands for "energy," *m* stands for "mass," and *c* stands for "the constant, the speed of light"; and since light travels at the tremendous speed of 186,282 miles per second, Einstein's equation states a very small amount of mass is required to create enormous energy.

Eddington surmised from the atomic weights of hydrogen and helium that a process called "atomic fusion" might form a helium atom from hydrogen atoms, with a loss of atomic weight of 0.7 percent. He was exploring a hunch. If his hunch was right, then that weight loss could account for an enormous energy release. Eddington calculated that if the mass of the Sun were pure hydrogen, then it could produce energy in this way for 10 billion years—a time span that fit the known age of Earth and the other planets in the solar system.

However, Eddington had not yet proven his point. Did the Sun have enough hydrogen to fuel this process for as long as he had estimated? That's where Payne-Gaposchkin stepped in and cleared up this part of the problem. She was able to show that hydrogen is the main ingredient of all stars, in spite of what seemed like unconvincing evidence.

A sound and unyielding researcher, Payne-Gaposchkin became the first person ever to receive a Ph.D. in astronomy from either Radcliffe or nearby Harvard (where she had taken courses). For her degree, she wrote a dissertation on the composition of the atmosphere of stars. At the time, astronomers were unsure about some strange results obtained when looking at stellar spectrograms (photographs or other representations of the spectrum of a star). There seemed to be a lot of unexplained variation. The most accepted explanation was that the variations represented differing amounts of elements found from one star to another.

Payne-Gaposchkin argued that, in fact, these results reflected differences in ionization, or temperature. Also, she found that the amounts of metals such as silicon and carbon occurred in about the same percentages on the Sun and other stars as on Earth. However, she suggested that the amounts of hydrogen and helium were vastly greater than on Earth. It took other astronomers some time to accept Payne-Gaposchkin's position, but they finally admitted she was right. Ultimately, it became clear that hydrogen and helium are the two most abundant elements in the universe—the stuff that stars are made of.

So, because the planets formed from the same solar nebula that formed the Sun, it followed logically that the atmosphere of early Earth would also have had a high hydrogen content, before the gas's extremely light molecules escaped into space. Hence Urey's hunch that hydrogen should be prevalent in the atmospheric mix in his experiment.

That's when a young graduate student in chemistry turned up who seemed the perfect candidate to work out the details and run the experiment. As Stanley Miller would later recount (Henahan, 1996):

> Urey gave a lecture in October of 1951 when I first arrived at [the University of] Chicago and suggested that someone do these experiments. So I went to him and said, "I'd like to do those experiments." The first thing he tried to do was talk me out of it. Then he realized I was determined. He said the problem was that it was really a very risky experiment and probably wouldn't work, and he was responsible that I get a degree in three years or so. So we agreed to give it six months or a year. If it worked out fine, if not, on to something else. As it turned out I got some results in a matter of weeks.

So, under Urey's direction, about the time that Francis Crick and James Watson were struggling to work out the double-helix arrangement of DNA, Stanley Miller set up a landmark experiment in the history of biology, an experiment aimed at simulating the imagined early-Earth scenario. He and Urey postulated a time when great clouds of gases rolled across the turbulent surface of the planet and lightning flashed across the skies—when gas molecules of methane, hydrogen, ammonia, and water were jolted into various molecular combinations by these bolts of lightning. A time when inorganic building-block molecules rained down into the Earth's shallow oceans. There, Urey and Miller imagined, these preorganic building blocks bumped into each other and eventually combined into longer, more complex, organic molecules, such as amino acids, proteins, and nucleotides. Ultimately, in this imagined sequence of events, these molecules would grow more and more complex until they developed into a nucleic acid capable of replicating itself.

In a small-scale version of this scenario, Miller created an atmosphere of hydrogen (H_2), with components of methane (CH_4) and ammonia (NH_3), floating over a flask of carefully sterilized and purified water. Into this primordial soup of gas and liquid, he introduced an electric charge to simulate lightning. For now, in their scenario, there would be no free oxygen, but plenty of hydrogen.

The results were astounding. Within a couple of days, Miller observed a pinkish-orange substance forming. After operating his experiment for a week, Miller noticed the water in the flask and in the trap below it had turned orange-red. He found, to his great excitement, that, in addition to simple substances, he had produced two of the simplest amino acids, plus some indication that a few more complicated ones were in the process of forming. Tests showed that the water contained amazingly high concentrations of amino acids.

When he conducted longer trials, more amino acids formed; other researchers tried the experiment and found they could repeat it. The experiment has been repeated successfully many times, producing all 20 protein-forming amino acids. The amazing thing was that the kind of organic molecules that formed in Miller's apparatus were the same kind that are found in living organisms. Miller had not created living creatures, but he had set in motion a process that seemed to be pointed in that direction. Maybe the development of life was not so unusual after all, but a natural consequence of the way the universe is constructed. In the late 1960s, this line of thinking gained even greater credibility when more and more complicated molecules were discovered in gas clouds in outer space.

Sagan was an undergraduate at Chicago when Miller completed his experiment. Sagan took a tour of the basement laboratory Miller called "the dungeon" and witnessed the conference at which fellow scientists posed skeptical, even scathing, questions. For Sagan, this was a view of the tough side of science, the political side in a sense, when the results put forth are not popular or not what people want to hear. Some of it was the justifiable need to establish that the experiment was well managed, that its process had no flaws. Some of the hostility had other, emotional sources. Sagan saw how hard this was on Miller. It was in some ways a glimpse into some of the difficulties—jealousies and prejudices—that Sagan himself would encounter later in his career. As of mid-2004, Miller continues to explore actively the many issues surrounding the origins of life and enjoyed a 50th-anniversary celebration of this experiment in 2002. But those early days required a tough skin for survival.

Later, during the 1960s, Sagan would build on the work of Urey and Miller. Sagan's method was similar, but the so-called primordial soup à la Sagan was a slightly different mix. Instead of methane, ammonia, water vapor, and hydrogen, Sagan followed updated estimates of early Earth's atmosphere, using methane, ammonia, water, and hydrogen sulfide. He simulated the effects of sunlight on these combined chemicals by exposing his mixture to ultraviolet (UV) radiation. Later in Earth's history, plants would begin photosynthesis and produce oxygen, which in turn would contribute to the formation of an ozone layer in the upper stratosphere, protecting the regions below from the Sun's UV rays. But in the beginning, Sagan reasoned, there must have been plenty of UV radiation to jump-start the biologic process. Sagan's experiments, like Miller's, also produced those fundamental building blocks of life, amino acids. Additionally, they produced several sugars and, most importantly, nucleic acids. These last are key to two fundamental life processes: the ability to pass on characteristics through genetics and the formation of proteins. Sagan's work, combined with the Urey/Miller experiments, provided considerable support for the idea that life could have formed through ordinary chemical and physical processes, without nonnaturalistic intervention.

LIFE ON OTHER WORLDS

Thus began Sagan's entrance into the field of "exobiology." The word was coined by Nobel laureate Joshua Lederberg to designate the study of life beyond Earth. But so far, no one has actually found proof of life on other worlds, a fact that encourages scoffers, who jibe that it is a subject without subject matter. However, the success of Miller's experiment shows that under the right conditions the key building blocks of life could have formed on prebiotic Earth. So the experiment offered substantial encouragement to the search for life on other planets. The reasoning went like this: If an experiment replicating a set of early planetary circumstances on Earth can produce materials key to life, isn't it logical that somewhere else in the billions of galaxies in the universe, life could (and would) happen? Why would we imagine that life would occur only here? By extension, if life could occur here, it could occur elsewhere.

Evidence reinforcing these ideas stacked even higher in 1970 when Sri Lanka–born biochemist Cyril Ponnamperuma (1923–1995) found traces of five amino acids in a meteorite that had landed in Australia on September 28, 1969. After careful analysis, he and his team of investigators found glycine, alanine, glutamic acid, valine, and proline—the first *extraterrestrial* constituents of life ever to be found. Ponnamperuma was able to show that the meteorite could not have picked up these amino acids through contamination by its contact with Earth. Rather, the amino acids were very likely synthesized by processes like those that took place in Miller's and Sagan's experiments.

As Ponnamperuma pointed out, the main goal of the research was to develop insights into how life began—on Earth and beyond that narrowly defined goal, how it may have begun throughout the universe. If the origins of life on Earth could be understood, he postulated, the same sequence of events might well have occurred in an enormous number of other host planetary systems as well.

More recent evidence has led most scientists to conclude the building blocks of life did not originate on Earth—at the time life began on Earth the atmosphere probably did not have the make-up envisioned by either Miller and Urey's or Sagan's experiments. Instead the building blocks for living things most likely came from outer space, carried by comets and meteorites. Close-up examination of comets such as Halley's Comet in 1986, as well as spacecraft flybys of some of the moons of Jupiter and other outer planets, show the presence of large quantities of dark, organic material, or "goo." Meanwhile, experimental results have shown that many of the complex building blocks of life (including amino acids) could survive a meteorite's fiery dive through Earth's atmosphere to the surface.

The Urey/Miller experiment and Sagan's later follow-up remain watershed experiments, however, showing the building blocks of life, including amino acids, can be formed from ordinary chemicals available in the universe.

UNIDENTIFIED BUT INTRIGUING

In 1996, Sagan would write in *The Demon-Haunted World*, "There are no cases—despite well over a million UFO reports since 1947—in which something so strange that it could only be an extraterrestrial spacecraft is reported so reliably that misapprehension, hoax, or hallucination can be reliably excluded" (p. 81). He liked to point out that no one would contest that UFOs (literally, "unidentified flying objects") are frequently spotted, but that doesn't mean they are spaceships from far-off stars. However, he adds, "There's still a part of me that says, 'Too bad'" (p. 81).

As that final comment suggests, Sagan had not always been such a severe critic of the UFO craze. He was a fan of science fiction and had not always held such a hardheaded view of the idea that visitors from other planets might be plying our skies. As a college student who had grown up on the fables of Burroughs, he had asked, Why not? He discussed his ideas with Muller, who listened good-humoredly to the younger scientist's ideas. Amidst a spate of front-page newspaper stories about UFO sightings, Sagan even wrote a letter during the summer of 1952 to Dean Acheson, who was secretary of state at the time, suggesting the government should perhaps take precautions in case alien visitors may not be friendly, and inquiring what defense plans were in place.

As late as 1954, Sagan was still defending UFOs to Muller, who perhaps teasingly suggested that his friends the Soviets were sending mysterious craft with astounding technological capabilities to hover over American soil. Sagan apparently took the remark seriously, refuting the concept point by point in a letter to Muller and defending the idea that the craft bore visitors from other worlds. It was his last known stand on that side of the issue. Soon thereafter, he changed his position to clear refutation, accompanied only by a wistful remnant of his former stance: "Too bad."

Everyone has committed some deed, made some remark, or held some view in the past that in retrospect seems foolish, and Sagan must surely have felt a little embarrassed at his overly zealous youthful defenses of the concept of alien spacecraft crisscrossing Earthly skies. As he would later point out, "Keeping an open mind is a virtue—but, as the space engineer James Oberg once said, not so open that your brains fall out" (Sagan, 1996, p. 187).

As he matured, Sagan became a solid, careful, reasoning thinker who would one day write about "my lifelong love affair with science" (Sagan, 1996, p. 28). Taking the position that extraordinary claims require especially tight evidence, he became a nonbeliever in UFOs. In fact he pointed out that people who defend reported sightings and the concept of alien spaceships always ask questions in terms of belief, or faith: Do you believe in UFOs? They almost never ask, Do you think the evidence for UFOs is convincing?

Sagan's ability to let go of a hypothesis in which he had invested considerable emotional energy is a sign of a good scientist. Science is a self-correcting process: When evidence does not fit a hypothesis, science calls for rethinking and revising the hypothesis, based upon the most solid evidence available. This is why, as improved methods and instruments for measurement and observation are developed, and as new information comes to light, scientific opinion changes. This self-correction is one of the most important gifts of science—and sometimes the most difficult for humans to accept. It requires a willingness to hold ideas tentatively, a readiness to change course if new, more convincing evidence becomes available.

This adjustable mindset frequently.conflicts with the less adjustable belief systems of religion. While some scientists accept the idea that religious thought functions quite differently than scientific thought, Sagan let go of religion as early as age 13, as he prepared for his bar mitzvah, the Jewish coming-of-age ceremony for boys. His careful study of biblical passages at that time led him to notice the many contradictions he found in the Bible, and he was bothered by the lack of consistency and logic. By the time he was in college, he had several vociferous arguments with his mother, who steadfastly held her religious beliefs and relied on them for comfort. Finally, though disappointed, Rachel Sagan realized she could not change her son's position on this issue. His position was consistent with his commitment to science and rational thinking, and it fit with the

Carl Sagan he was in the process of creating, a person committed to the nonmiraculous nature of the world and life.

LYNN

During his college years, Sagan met fellow University of Chicago student Lynn Alexander, whose role in Carl's life was the first to compete with his mother's in importance. The two students met when they nearly collided on a stairway in the math building, Ekhart Hall, and struck up a conversation. It was the beginning of an important relationship for Carl. Sagan was 20, and Lynn Alexander was 16. Intellectually precocious, she was already a veteran student at Chicago, having escaped her local high school two years earlier for the more challenging university classes. Not easily swept off her feet, she nonetheless thought Carl was very interesting, with an amazing command of language and tremendous confidence. Later, when Lynn had become a distinguished, renowned scientist (now bearing the name Lynn Margulis), she would recall her impression of Carl as "tall, handsome, with a shock of brown-black hair, and exceedingly articulate, even then... full of ideas.... His love of science was contagious" (Davidson, 1999, p. 67). They soon became part of each other's lives.

Sagan's friends liked Lynn. They appreciated her sense of humor, her eager delight in the world of ideas, and her fascination with the world of science. They found her intense, open-minded, lively, and most of all, warmhearted and exceptionally intelligent. She was also very attractive and full of questions about biology in general and genetics in particular. She and Carl talked nonstop about this shared interest, and he was delighted to have an audience to listen to his view of life as a cosmic presence rather than a unique feature of Earth. He also told Lynn he believed scientists would ultimately explain life-forms solely in terms of physics and chemistry.

Lynn's view of biology, as it developed over the coming years, was equally visionary, but she saw life as Earth-specific; she saw Earth and all its living inhabitants as a single, unified organism. In the 1970s, she and British scientist James Lovelock would put forth the "Gaia" hypothesis, named after the Greek goddess of the Earth, based on these tenets.

Lynn's and Carl's views were incompatible—or at least the two young scientists never found a way to reconcile them. Lynn tended to think Carl's quest for extraterrestrial life was silly and pointless. Her Gaia vision made Carl uneasy. Their shared interest, which might have formed a solid bond between them, would instead become a wedge between them, or at least an ever-present thorn. Both were ambitious and focused on their careers and neither needed the distraction and frustration of an adamant naysayer so near at hand.

Perhaps the relationship was doomed at the start, but both were attracted by the pluses, and the minuses took several years to accumulate. In the meantime, they had both good times and not-so-good times, as most young couples do.

BATHING WITH ICE

Despite his ability to breeze through essay exams and routinely ace term papers, not everything went smoothly in Sagan's academic life. His high grades placed him in an honors program that required the completion of a thesis. Because of his interest in the Urey/Miller experiment, he chose H. C. Urey as his adviser and proceeded to write his thesis on the origins of life. With his usual confidence, he submitted it to Urey for review. However, when the paper came back, it sent a chill down Sagan's spine. It was covered with copious notes painfully pointing out error after error. Sagan would later remark he felt as if he had taken a sudden "plunge into an ice bath" (Poundstone, 1999, p. 25).

With considerable care, Sagan rewrote the paper. It was accepted, and in 1954, after three years of study, he received what he laughingly called his A.B. degree "in nothing" from the University of Chicago. According to Chicago's unusual curriculum, students were allowed to declare a major *after* they received an A.B. degree. The following year, Sagan declared physics as his major and earned a second bachelor's degree. In 1956, he followed up with a master's in physics as well. (Given his varied interests, he might instead have chosen biology, chemistry, genetics, or astronomy. But physics was a sound, fundamental discipline, a strong background for an astronomer and a choice in keeping with his premise that biology would ultimately be seen in terms of chemistry and physics.)

He was also admitted to the University of Chicago's Ph.D. program in astronomy, which was taught out of the Yerkes Observatory in Williams Bay, Wisconsin. He would start in the fall. Meanwhile, the observatory director, Gerard Kuiper, asked Sagan to join him for summer observations of Mars at the McDonald Observatory in Fort Davis, Texas. Mars would be at opposition (where it can be seen more easily than at any other point in its orbit). For Sagan, it seemed like the chance of a lifetime—observing Mars, his favorite planet, with the only widely respected planetary astronomer in the country. So after commencement, Sagan eagerly headed for Texas.

Chapter 3

ASCENT OF A PLANETARY SCIENTIST

McDonald Observatory in Fort Davis, Texas, is operated by both the University of Chicago and the University of Texas, and Dutch-born American astronomer Gerard Peter Kuiper (1905–73) directed both the Yerkes and McDonald observatories from 1947–60. Kuiper was a giant in solar system research and observation, and the opportunity to observe Mars at his side was one that would have been coveted by any aspiring planetary scientist. Sagan set off for Texas with high hopes, as Lynn Alexander also headed south—a little farther south, though, to work on a prestigious anthropology project in Mexico.

MARS VISIBILITY: POOR

Kuiper, who earned his Ph.D. at Leiden University in 1933, spent three years as a fellow at Lick Observatory on Mount Hamilton in California and joined the astronomy faculty at the University of Chicago in 1936. He became a U.S. citizen in 1937.

Kuiper brought a new focus to solar system astronomy with his discoveries using McDonald Observatory's big 82-inch reflecting telescope. In 1944 he determined the composition of the atmosphere of Titan, the largest moon of Saturn—currently considered one of the most intriguing objects in the solar system because of its unusually thick atmosphere containing methane. In 1948, he discovered carbon dioxide in the atmosphere of Mars, and he also reported, based on infrared observations, that the polar caps are ice, not frozen carbon dioxide as some scientists thought. This observation caused ongoing controversy, but finally received validation by direct observations made in 2004 by the European spacecraft *Mars Express*. In the late 1940s Kuiper also found Miranda, a small moon close to Uranus, and Nereid, a small moon of Neptune. In 1951 he proposed that the planets and their moons formed independently by condensation from gaseous protoplanets, in contrast to the prevailing theories of more cataclysmic origins. Kuiper also made observations and measurements of Pluto and suggested that other small bodies orbit the Sun in a belt beyond the orbits of Neptune and Pluto. Today, this region is known as the Kuiper Belt, in his honor. With the dawn of the space age in the late 1950s, the solar system became even more interesting to the public. It was the perfect time to begin a career in planetary science, and who better to study under than the master himself?

The observation of Mars during the summer of 1956 was anticipated with great eagerness worldwide, and internationally coordinated efforts took place especially in the Southern Hemisphere, where it could be seen best. Mars would be in opposition, which meant the Sun, the Earth, and Mars would line up, with the Sun and Mars sandwiching the Earth in between, bringing Mars unusually close to Earth, only some 40 million miles away (compared to a maximum distance of 249.4 million miles). It was the closest Mars had come to Earth in 79 years, and the planet's red glow was expected to hang brightly in the night sky. But the summer of useful observations was not to be for Sagan and Kuiper. Dust storms on both the surface of Mars and the plains of Texas made seeing cloudy at best. Kuiper could only report the presence of an enormous yellow cloud, some three thousand miles long, obscuring surface features. And Sagan and Kuiper spent most of their time talking instead.

Kuiper had stories to tell. During World War II he helped the Allied military with technical expertise, and in the process he got hold of some infrared sensors used by soldiers for night fighting. He later used them to detect infrared radiation given off by astronomical objects.

Kuiper was also the source of a trick Sagan learned early on to evaluate the likelihood of a hypothesis. The highly useful order-of-magnitude technique involved sketching out the hypothesis in its simplest possible form, say on the back of a napkin, and roughly working it out. If the results looked plausible, then the hypothesis might be worth pursuing in more detail. If not, back to the drawing board—because something was probably wrong.

Kuiper and Sagan had a lot in common, especially their shared curiosity about the possibility of life on other planets. Like Sagan, Kuiper had postulated the existence of some form of life on Mars. Kuiper had seen his name splashed across the newspapers with news of his hypothesis that lichen-like chlorophyll-bearing plant life might cause the dark patterns that spread and receded across Mars with the seasons. In the world of astronomy, this sort of attention is not always greeted with approval from colleagues, as Sagan would also soon learn. It tends to strike the scientific community as hype. Not many years earlier, astronomer Percival Lowell mistakenly reported observing canals on the surface of Mars. The press and even other astronomers were so taken with the idea that it got out of hand. Newspaper reporters let their imaginations run wild, reporting evidence of a vast, intelligent, canal-building civilization on Mars. Better telescopes did not confirm the observation, however, and the embarrassing brouhaha was chalked up to an optical illusion. Astronomers were still smarting over the widely publicized mistake. But to Kuiper, the idea of life on another planet still seemed plausible as it also always did to Sagan—even if only in the form of fossil remains of microscopic organisms.

Unknown to Kuiper and Sagan, that summer they spent with McDonald's big 82-inch reflector telescope stood close to the end of an era. On October 4, 1957, with the launch of the Soviet satellite *Sputnik*, humankind would break through Earth's atmosphere. It was a watershed moment in human evolution.

As the tiny, lonely, blinking sphere cut its path across the skies, its political significance was not lost on Americans. It was a gauntlet thrown down in the Cold War one-upmanship of the 1950s. A space race had begun between the United States and its ideological adversary, the Union of Soviet Socialist Republics (USSR). At the same time, *Sputnik* was a wonder to behold. Families went out into the autumn evening and watched the skies together to witness the tiny light of the small spacecraft's crossing. With it, humankind had crossed into the space age.

Sputnik II followed on November 3. The Soviet Union—ideological foe and political adversary of the United States—reached space first with the launch of the two *Sputniks*. To catch up and save face, the United States marshaled the expertise of the U.S. Navy, Army, and Air Force, including Wernher von Braun and his rocket experts. By January 31, 1958, the United States followed with its own satellite, *Explorer I*. The civilian-led National Aeronautics and Space Administration (NASA) was born the following autumn.

The significance for astronomers was huge. At last, they could reach the regions outside Earth's atmosphere and peer at the universe from far above the hindrance of clouds, urban lights, and the atmospheric disturbances that plague ground-based observatories. Astronomers had been waiting for this moment, and they began planning the first space telescopes almost immediately. Within two decades, they would find an additional way to escape most of the water vapor that blocks reception of infrared radiation (given off by cooler objects in the universe)—by flying an infrared observatory aboard a customized high-altitude C-141 U.S. Air Force jet. It was named, appropriately, the *Kuiper Airborne Observatory*, and it carried astronomers high above clouds and vapor for more than 20 years, from 1974–1995. All these efforts were the forerunners of such stunning astronomical successes as the *Hubble Space Telescope* (launched April 25, 1990), the *Chandra X-ray Telescope* (launched July 23, 1999), and the *Spitzer Space Telescope* (launched August 25, 2003). So the events of 1957–58 would bring astronomy to the verge of breaking out of its prison beneath the turbulent atmosphere of Earth—and within a few decades, robot spacecraft would bounce to rest on Mars, carrying robot rovers to explore its surface. And in 1956, Carl Sagan was in the perfect place at the perfect time—this time around for the reawakening of planetary astronomy.

At the end of Sagan's stay in Texas, before the fall 1956 term began at Chicago, he drove down to Mexico to visit Lynn. But the excursion did not go well. The spicy food aggravated his achalasia and the malted shakes he usually drank to make food slide down his throat were nowhere to be found. Hungry and in constant pain during the entire visit, he grew uncharacteristically grouchy. As his widow Ann Druyan puts it, "In Mexico, [Carl] was in agony" (Druyan, 2004). As a result, he was so rude to Lynn's colleagues there that she became embarrassed and angry. The situation was awkward, to say the least. Lynn felt he showed an extreme lack of sensitivity, intentionally or not, and she began to rethink their relationship. It was an unpleasant ending to the summer.

Looking back, one can easily see the incident was an anomaly. Sagan was known even then for his liberal outlook, and bigotry was alien to his character and background. Yet the pain, combined perhaps with the self-absorption that sometimes comes with youth, clearly blinded him to the impression he was making. He may have failed to recognize when he was being impolite. And at that stage Lynn did not have the fuller picture produced by his more empathetic behavior in later life.

YERKES

With the arrival of the fall term, Sagan began his studies at the venerable Yerkes Observatory. Astronomical observations first began at Yerkes in 1897, when it was founded by astronomer George Ellery Hale as the first combined observatory and research institute. Designed by Henry Ives Cobb under Hale's direction, the observatory is housed in a large Romanesque structure designed in the shape of a huge Latin cross and built of brown Roman brick. Ornate nineteenth-century figures dominate the building's columns. Two observation towers occupy the north and south extremities. The large dome at the western end of the building houses the big 40-inch refractor, which is still in use. Its great lens was ground by the legendary telescope-maker Alvan Clark of Massachusetts, and the telescope was completed in 1895. It remains the largest refractor ever built.

Yerkes is situated on the shore of Lake Geneva on the west side of the village of Williams Bay, Wisconsin. Just 78 miles across the border from Chicago, the town was erected on land once occupied by the Potawatomi Indians and their famous leader, Chief Bigfoot. The village was named after its founder, Captain Israel Williams, a Connecticut veteran of the War of 1812 who had wandered west in search of a new start in life.

The quiet, startling beauty of the village and its surrounding area quickly attracted wealthy citizens from Chicago, and by the 1860s, impressive yachts could be seen lazily cruising past expensive summer homes on the lake's quiet shores. The great Chicago fire of 1871 found many other wealthy citizens building summer homes in the area as they waited for their devastated Chicago mansions to be rebuilt.

Not without a bit of culture, the village's Williams Bay Repertory Company had given a start to the young actor Paul Newman, who began his acting career there in 1950, just six years before Sagan arrived to begin his doctoral studies at Yerkes.

The observatory continues to be a magnet for some of the finest astronomers and astrophysicists in the world. At the time Sagan began work there on his Ph.D., it was home to the brilliant astrophysicist Subrahmanyan Chandrasekhar, who would share the Nobel Prize in physics in 1983 for his work on stars and their structure. Chandrasekhar was aloof and dauntingly self-assured, and, even Sagan's typical self-confidence failed him before "Chandra" (as everyone called him). Sagan did take several classes from him, though.

Sagan's personal life during this time was turbulent. He and Lynn Alexander broke up, got back together, and broke up again, over and over. They had gone through phases of dating other people. They argued. And then all would be well for a time. One of the enduring issues was Carl's mother, Rachel. During the spring break of 1955, for example, when Carl and Lynn were both undergraduates, they traveled to New Jersey in Carl's blue-and-white Nash-Hudson station wagon. Lynn wanted to visit Princeton, which she did, and she visited J. Robert Oppenheimer, who directed the making of the atomic bomb during the years of the top-secret Manhattan Project. (She had written a term paper on the subject, "Not

'Whether or Not?' but 'How?': J.R. Oppenheimer and the Decision to Drop the Bomb" and was determined to meet him.) Meanwhile, Sagan visited with his mother. He did not invite Lynn, however. No proud introduction to Sagan's mother took place. Instead, Carl dropped Lynn off a few blocks from his home, fearful of upsetting Rachel.

Lynn would later recall Rachel's disapproval of their relationship, how she would ask Carl why he wanted to marry "that *scientist*" (Poundstone, 1999, p. 35). According to Sagan's sister, Cari Sagan Greene, "Rachel would never have thought *any* woman was good enough to be Carl's wife" (Davidson, 1999, p. 81). When Rachel heard about the wedding plans she went on an hours-long crying jag. Lynn was Jewish, which seems to have been her only plus as far as Rachel was concerned. So Lynn had more than a few misgivings about marrying Carl, many of them surrounding his mother and Carl's dependence on her. Then, after the incident in Mexico, Lynn became determined *not* to marry Carl, telling herself—on a tape recording—that marriage to Carl would be self-destructive. But the pluses finally outweighed the minuses in Lynn's mind, and she went forward with the wedding plans. Carl and Lynn decided to marry the following summer.

WEDDING BELLS AND PH.D.S FOR TWO

Carl Sagan and Lynn Alexander married on June 16, 1957, with Rabbi Jacob Weinstein conducting the ceremony. Carl envisioned a large family of brilliant children and thought they should start early. He understood (and supported) Lynn's career plans, so he sketched out a timetable. They would have five children at a rate of one every 18 months, and within just a few years, Lynn would be able to return to the process of earning her Ph.D., devoting herself full time to a career in research and teaching. By today's standards, this plan seems arrogant and self-centered, and no doubt Sagan never considered sharing the household and child-raising responsibilities. He should have, especially considering that Lynn was in her own right a brilliant scientist with fine potential. However, it is unfair to judge one period in time by the values of another. Sagan exhibited sexist attitudes that were common at the time, but, to his credit, he gave more thought than the average 1950s husband to his wife's goals outside the home. Carl actively encouraged Lynn's career and wanted her to pursue it. He just never considered how unfair his lack of participation in household management really was.

But the symbiosis of their relationship—a partnership for mutual benefit—worked well for them both. Lynn still saw and appreciated in Carl the characteristics that originally attracted her—his intelligence, his interest in ideas, his tall, lanky, handsome appearance, his ambition, and his love of science. He taught her what he knew about biology and genetics and provided her with important contacts in these fields. She gave him intelligent companionship, a sympathetic friend who shared with him a set of major intellectual commitments: the importance of science as a way of knowing, as a way of finding out about the universe, as a naturalistic approach to life, and as a release from the bonds of superstition.

A month before their wedding, Lynn received her bachelor's degree (like Sagan, she received a liberal arts degree with no declared major). The young couple took off to spend the summer at the University of Colorado in Boulder, where they lucked into a house with a gorgeous view, rented from a cousin of Sagan's who was away for the summer. At Colorado, Lynn studied ecology and plant physiology. Carl studied with distinguished Russian-American theoretical physicist George Gamow. In addition to work on cosmogony and radiation, Gamow had worked in astrophysics on the theory of the evolutionary universe. He had also written several popular books on science, including *The Birth and Death of the Sun* (1940) and *One, Two, Three . . . Infinity* (1947). These interests dovetailed nicely with Sagan's evolving interests, and Sagan gained intellectually from working with Gamow.

In the fall, Carl returned to his Ph.D. work at Yerkes while Lynn entered the Ph.D. program in zoology and genetics at the University of Wisconsin at Madison. They found a place to live in Madison, where big-city life offered more cultural activities and a cosmopolitan atmosphere than did Williams Bay or anywhere in between. The commute took Sagan about ninety minutes each way.

During the summer of 1958, while working in the biology lab, Lynn fainted. At that moment she knew that within fewer than nine months she and Carl would have their first child. The schedule had begun. The financial timing was poor, however. Carl's string of end-to-end scholarships, which had steadily supported him since 1951, had just ended, and Carl had taken part-time work in Madison. Lynn's fellowship would remain on hold until the baby was born. Somehow, though, they managed.

The baby was a boy, born on March 17, 1959, and Lynn and Carl named him Dorion Solomon Sagan, his first name partly given in homage to Irish-born writer Oscar Wilde's novel, *The Picture of Dorian Gray* (1891) and partly as an allusion to the constellation Orion (D'Orion, or "of Orion"). His middle name was a salute to their Jewish heritage.

Meanwhile, during this period Carl's career began to evolve. The year 1957 saw publication of his first peer-reviewed article, "Radiation and the Origin of the Gene," in the journal *Evolution*. But three other important career-building events also took place: He developed and presented an insightful alternative hypothesis to Kuiper's belief that vegetation existed on Mars; he began work on a groundbreaking hypothesis explaining high temperatures on the surface of Venus; and he met a key mentor who proved both willing and able to open many useful doors.

LICHEN? OR DUST STORMS?

In December 1956, following their summer in Texas trying to observe Mars, Sagan and Kuiper set off for a meeting of the American Association for the Advancement of Science (AAAS) in New York. Both had prepared papers to present. Kuiper explored his hypothesis that the waning and waxing of dark areas might be lichen-like organisms responding to seasonal changes. Sagan put forth a different explanation.

Actually, Sagan pointed out, observers using telescopes had found no link between the seasons and either the size or location of the changing dark regions. He wondered why. Could giant dust storms (like those of the previous summer) blow across areas of dark lava, making the darkness disappear, only to reappear when another windstorm swept away the light-colored dust? Thus the dark regions would seem to regenerate, then die off, then regenerate anew. This explanation fit with the recorded observations, and it had the advantage of simplicity. So Sagan's explanation would be favored under the principle of Occam's razor (the principle of economy in logic, attributed to the fourteenth-century philosopher William of Occam, which suggests the simpler of two possible explanations is usually closer to the truth). Sagan's hypothesis—which turned out to be correct—was especially interesting because he *wanted* to find living organisms on Mars. Also, he was going against Kuiper's pet hypothesis, which was courageous, to say the least. Ironically, the reporter for the *New York Times* got it all wrong, crediting Kuiper with Sagan's ideas, and Sagan with Kuiper's (Poundstone, 1999, p. 35).

TEMPERATURE: HELLISH

Sagan was also interested in Venus, long thought to be Earth's twin, about the same size and distance from the Sun, and blessed with an ample atmosphere. Science-fiction writers capitalized on the absence of detail known about the surface of Venus, which no one had ever seen. Perhaps there were lavish steamy jungles inhabited by strange and wonderful creatures. Even scientists joined in. Nineteenth-century Swedish chemist Svante August Arrhenius suggested the Venusian surface was soppy and wet, its surface bogged with swamplands. Aside from the fact astronomers had not been able to detect any water in Venus's atmosphere, no one had any evidence to the contrary. The veil of Venus never lifted.

What astronomers *could* see was carbon dioxide—lots of it—in the atmosphere. Harold Urey took the chemist's view: Why would carbon dioxide show up and not water? On Earth, reactions among carbon dioxide, water, and silicates form carbonate rocks and silica. He saw no signs of any of these reactions. But what if there was no water on Venus? Then no such reactions would take place. Venus, said Urey, was probably bone dry.

Meanwhile, Fred Whipple and Donald Menzel of Harvard had an entirely different idea. What if all land on Venus were covered with vast oceans. This would be another scenario in which no reactions would take place among carbon dioxide, water, and silicate rocks—because no such rocks would be exposed to both compounds.

In a third corner stood Fred Hoyle, an astronomer (and science-fiction writer) from the University of Cambridge in England. He saw Venus as home to gooey oceans of petroleum hidden by a cloud cover of smog. In the beginning, according to his view, Venus started out with a great preponderance of hydrocarbons and very little water. Reactions took place over time between the carbon in the hydrocarbons and the oxygen in the water, producing vast quantities of carbon dioxide and exhausting the water supply.

But Sagan noticed a supply of new data that fit none of these scenarios. A group of scientists using a radio telescope on the roof of the Naval Research Laboratory (NRL) obtained a very strong reading from Venus in the microwave region of the electromagnetic spectrum. That meant Venus (like a microwave oven) was emitting heat—lots of it. By analyzing the spectrum produced by these microwave emissions, it was possible to estimate the temperature of the object producing the radiation. The estimate was 600 degrees Fahrenheit. This news was generally resisted by the scientific community. It was hard to let go of the "Earth's twin" vision of Venus, and to give up the ever-present hope of friendly, intelligent neighbors going about their daily lives just across the ocean of space between the two planets. How could there be such incredibly high heat scorching the surface of our next-door neighbor? Alternate readings of about 110 degrees Fahrenheit were more welcome. Probably, argued some scientists, the hotter reading came from a hot spot in the ionosphere.

Sagan suggested something simpler: A greenhouse effect. Decades earlier, Arrhenius had already studied this effect on Earth. The process was well understood. Like glass in a greenhouse, carbon dioxide in the atmosphere admits light from outside but blocks the infrared radiation thus generated from passing back through the carbon dioxide layer to be radiated out into space. So the surface becomes warmer and radiates heat into the atmosphere. Some greenhouse effect can be beneficial to life, as it has been for eons on Earth. But Venus could be another story. Sagan did some calculations. He found that a greenhouse effect could account for some of the heat on Venus, but not all of it. Carbon dioxide allows selected frequencies of infrared radiation to pass through. But carbon dioxide is not the only greenhouse mechanism. Water vapor also blocks the escape of infrared radiation. Sagan did some more calculations. Water vapor would block heat from escaping at the very frequencies that carbon dioxide missed. This scenario, Sagan deduced, could cause a mighty greenhouse effect, which in turn could account for the high temperature readings obtained by the team at the NRL.

No evidence for the presence of water vapor in the Venusian atmosphere existed at the time. Yet the hypothesis made sense. Sagan made his mark, and his position was later vindicated by measurements made by robot spacecraft. The surface of Venus was extraordinarily hot (600 degrees Fahrenheit was conservative, 864 degrees Fahrenheit is more accurate). Life in any known form would be impossible.

LEDERBERG, NASA, AND BERKELEY

If the Sagans had not decided to live in Madison, Carl might never have met his most important mentor, geneticist Joshua Lederberg, cowinner of the Nobel Prize in physiology or medicine in 1958 (which is probably the year the two men met). Back in 1944, Lederberg had discovered that bacteria have a sex life, of sorts. That is, they reproduce by conjugation, or the mutual exchange of genes between single-celled organisms that have no sex differentiation. Perhaps even more important, he later found some viruses can carry genetic information from one bacterium to another, and in the process they modify the inherited characteristics of host cells. This is the work for which Lederberg won the Nobel. Moreover, he and Sagan had a mutual interest in the concept that life might exist on other worlds—in fact, Lederberg coined the word for this concept: "exobiology." (The expectation initially was that life would be found on other planets in the *solar system*—and that may still happen, although the odds seem much lower now than in 1958. So exobiology came to imply, in some circles, an interest in how life might exist elsewhere in the solar system. "Astrobiology" is the term used today to imply wider applications—biological organisms elsewhere in the *uni-*

verse.) The synergy between the two scientists was instantly present when they met—despite 33-year-old Lederberg's reputation for tearing graduate students' arguments to pieces.

Lederberg had begun to think the Urey/Miller experiment made an unnecessary assumption: that life began on Earth. What if it had originated elsewhere—even in deep space? Primitive life could have been carried to Earth by comets or asteroids, raining down upon Earth's oceans. As early as the nineteenth century scientists had considered an idea they called "panspermia," which set forth just such a hypothesis. British geneticist and writer J.B.S. Haldane, a friend of Lederberg's, revisited the concept in 1954.

Lederberg thought the idea held merit and began pondering its implications for the Moon—which would also have received its share of raining "astro-plankton," as Haldane called these alien microbes, or at least the Moon might have received showers of extraterrestrial organic molecules. With Dean B. Cowie, Lederberg coauthored a paper entitled "Moondust," published in *Science* in 1958. In "Moondust," they suggested layers of frozen dust and organic molecules would have accumulated over billions of years of this constant bombardment, so that should astronauts from Earth set foot on the Moon, they would encounter deep piles of organic material for biologists to examine. Examination of the Moon's airless, minimally changing, deep-freeze environment might lead scientists to a new understanding of how life formed on Earth. These ideas fit neatly with some thinking Sagan had done about organic molecules in space. Lederberg knew biology, and Sagan knew planetary science, so each was able to fill in the gaps for the other, and their partnership—trading books and ideas back and forth—lasted some ten years. Each was open to different points of view, and each learned from the other.

Lederberg often defended Sagan against colleagues who criticized Carl's thinking as bizarre and off-the-wall. "He had a lot of offbeat ideas," Lederberg later conceded. "They were always at some level *not* illogical, and some of them could prove to be right; and I would point out [to others] the value of listening closely to someone who has that degree of rigor and imagination at the same time. [On] any topic he got into, he certainly did his homework—very, very thoroughly" (Davidson, 1999, p. 90). Sagan believed deeply in the application of rigor and discipline to the scientific process, and, because he looked closely at the evidence, he often developed and defended positions running counter to his own hope of finding life on other planets. (Such was the case with his ideas about dust storms on Mars and the surface temperature on Venus.)

But Sagan also threw a very wide net—gathering up information from diverse disciplines and integrating ideas. Early in his career, especially, he may sometimes have erred in this process because he did not always delve deeply into any one aspect of a question. This, coupled with his verbal facility, sometimes gave an impression of skimming the surface, when actually he had synthesized a great deal of information.

Meanwhile, Lederberg, who organized the Department of Medical Genetics at Madison, and served as chair from 1957–58, headed to Stanford the following year to organize a medical genetics department for that prestigious university.

Lederberg's speculations about the Moon and its possible interests to science made him very particular on the subject of biological contamination—the process of despoiling the Moon's native biology by introducing alien biotics from Earth, or vice versa (back-contamination, the fear that microbes from the Moon might travel back to Earth). So, after *Sputnik* and the recognition that spacecraft might soon barrage the surface of the Moon (the first successful robotic landing was made by the USSR in 1959), he actively made the case for preventing contamination. He encouraged concern among members of the governing council of the National Academy of Sciences (NAS). Lederberg's concern about planetary contamination was also well received by NASA. As the new agency began to take shape, its administrator (and highest official) Hugh Dryden set up a Space Sciences Board within the NAS. Lederberg headed the exobiology panel of that board and helped start discussion groups about the topic on both the West and East Coasts. These groups were composed of the best minds in the land—including Harold Urey, Stanley Miller, and Melvin Calvin. In a letter to NASA's Robert Jastrow, Lederberg suggested NASA hire Sagan as a part-time researcher and adviser for the western group—to advise the group based on his review of the literature on subjects they were discussing, to prepare their reports for publication, and to write a "Handbook of Planetary Biology." Suggested part-time salary: $4,000 a year. NASA complied—putting Carl, still a graduate student, on a footing with Nobel laureates and other luminaries. Moreover, by the autumn of 1959, Sagan would publish his first article on the subject of planetary contamination in the AAAS's highly selective journal, *Science*.

Getting to know Calvin worked out well, too, for Sagan, since they shared an intense interest in the origins of life and life on other worlds. In particular, Calvin was interested in evidence of life *from* other worlds, including the possibility of finding organic molecules in meteorites—space rocks that have landed on Earth. Calvin, who would receive the Nobel Prize in chemistry in 1961, respected Sagan's contributions to the contamination discussions and asked Sagan to spend the summer of 1959 researching in his lab at the University of California at Berkeley. This opportunity gave Carl hands-on experience with an infrared spectrometer—a fundamental instrument in a planetary scientist's tool kit. That summer would also lay the groundwork for a postdoctoral fellowship not far down the road.

JOVIAN CONJECTURE

As part of his doctoral research, Sagan examined the highly speculative possibility of life on Jupiter, which led him down yet another path. Despite the very cold temperatures of Jupiter and the crushing pressure of the giant planet's atmosphere, Sagan postulated Jupiter might duplicate on a huge scale some of the atmospheric chemistry in Stanley Miller's experiment, supposing (as most astronomers did) that Jupiter's atmosphere was composed primarily of methane, ammonia, water, and hydrogen (the same ingredients Miller used in his experiment). Carl paid a visit to Miller at Columbia University in New York in 1959. Would Miller be willing to run some experiments for him, simulating the conditions he thought might exist on Jupiter? Miller agreed—somewhat nonplussed by the request, but won over by Sagan's persuasive charm. Miller ran the experiments and shipped the results to Sagan, who by this time was researching in Calvin's lab at Berkeley. Sagan used the infrared spectrometer to analyze the chemical composition of Miller's samples. Sagan reasoned the atmosphere of Jupiter, a known source of radio noise, might be subject to intense electrical storms, especially in the turbulent region of the Great Red Spot. Lightning bolts passing through the atmosphere might result in brightly colored, complex organic molecules, thereby explaining the intense coloration of the giant planet. If so, could such a concoction of organic molecules result in a form of life indigenous to Jupiter, floating in the atmosphere's organic-laden cloud layers?

In the long run, Sagan's experience with the infrared spectrometer and his relationship with Calvin and his lab would prove more important than his exploration of this subject, which he always recognized as only remotely possible. When he presented a paper on the subject to the Radiation Research Society in San Francisco the following May, he told a press conference, "Look, I'm talking about organic molecules, not life. I'm not saying that there's life on Jupiter. I'd be very unhappy if any of you wrote that I said, 'life on Jupiter.'" Sure enough, the next day at least one San Francisco newspaper ran the story as "Life on Jupiter, scientist says" (Poundstone, 1999, p. 44).

LEAVING YERKES

Sagan received his Ph.D. in astronomy and astrophysics from the University of Chicago in 1960. Typical of the wide net Sagan liked to throw, his dissertation on "Physical Studies of the Planets" covered several subjects: (1) "Indigenous Organic Matter on the Moon," (2) "Biological Contamination of the Moon," (3) "The Radiation Balance of Venus," and (4) "Production of Organic Molecules in Planetary Atmospheres: A Preliminary Report." The unstated common theme: Life on other planets.

The first of these—the possibility of organisms or building blocks of life on the Moon—dovetailed with the concern about biological contamination he shared with Lederberg and others in his NAS discussion group.

The third topic had already led him to correspond with an interesting and capable radio astronomer named Frank Drake, a contact that would lead Sagan to another chapter of his career not far down the road. Meanwhile, Sagan's dissertation work on Venus met with critical respect and acclaim, and his article "The Planet Venus" was published in *Science*, March 24, 1961.

As Sagan's time at Yerkes wound down, he turned to the question of his postdoctoral work. In the waning days of 1959, he had applied for a National Science Foundation (NSF) grant to study in Paris at the Meudon Observatory, where he could work with French planetary scientist Audouin Dollfus. Meanwhile, though, Lederberg had recommended him for the two-year Miller Fellowship at the University of California, Berkeley. The Berkeley faculty was a powerhouse—composed, in addition to Calvin, of such giants as Emilio Segrè, Luis Alvarez, Glenn Seaborg, Edwin McMillan, and Owen Chamberlain. The West Coast was also fast becoming a center for space science. These factors, combined with the honor, length, and stipend of the fellowship, made it a real plum. Calvin, Lederberg, and Kuiper all wrote strong recommendations for the Miller. Sagan received both awards. He had the incredible luxury of choosing, and based on advice from Kuiper, Sagan decided to accept the Miller fellowship. Lynn, who was working on her master's degree in zoology and genetics, was admitted to the graduate program at Berkeley. Carl, Lynn (who was expecting their second child), and Dorion were headed for California.

Chapter 4

COME IN, UNIVERSE

Carl and Lynn found a house high in the Berkeley hills overlooking the Golden Gate Bridge and the San Francisco Bay. Sagan's 1959 fellowship income of $7,500 a year (the equivalent of some $46,600 in 2003 dollars), augmented by $500 a year for contingencies—such as travel, supplies, and equipment—in addition to his income from his NAS post put them in a good financial position for so early in their careers.

Carl settled easily into life in California, where the weather was warmer, eccentricities were tolerated and even encouraged, and the drive home wound through beautiful hills bathed gold in the setting sun.

BERKELEY

As a Miller Research Fellow, Sagan was officially attached to the UC Berkeley Astronomy Department, Institute for Basic Research in Science, and Space Sciences Laboratory. The U.S. space program was growing rapidly—primarily because of the role that space exploration played in the Cold War with the USSR after the launch of *Sputnik*. Showing off U.S. technical expertise and strength on the world stage could make or break the nation's position in the international pecking order. Enthusiastic funding for space meant healthy windfalls for engineering and science, especially space science. Exobiology—Joshua Lederberg's favorite topic—in particular attracted generous funding. Now, with Sagan at Berkeley and Lederberg across the bay at Stanford, the two scientists could resume their practice of bouncing ideas off each other, comparing notes, and exchanging resources. Sagan was interested in every aspect of exobiology. He continued his interest in producing scenarios that might simulate the origins of life. Like Melvin Calvin, he also followed the news regarding organic molecules found in meteorites. And he found intriguing a project he heard about in the new field of radio astronomy: "Listening" for possible messages from alien civilizations, based on the idea we probably are not the only intelligent life in the universe.

IS ANYBODY OUT THERE?

With the invention of the optical telescope in Galileo's time, people had begun to realize that planets were worlds something like our own. And they also began to wonder if anybody lived there. Recognizing that our Sun is a star like many others in the sky, they wondered, too, if some of those other stars might also have planets orbiting dependably about them and if those planets might also hold civilizations like ours. Or these alien civilizations might be more advanced, or less so. We might be able to learn from them, or teach them. This is the stuff from which good science-fiction stories could be spun. (And many were.) But with the coming of the space age, these thoughts, like space exploration itself, entered the realm of serious scientific investigation. Enter a lone scientist working at the newly founded National Radio Astronomy Observatory in Green Bank, West Virginia.

Frank Drake (1930–) was born in Chicago where he was raised on the South Shore. He became interested early in all aspects of science—from how motors worked to the size of the universe—and he and another boy from his neighborhood used to ride their bikes to the museum to learn more. He began wondering, even at that early age, about whether there were other worlds orbiting around other stars, average stars like our Sun. He began to think "perhaps humanity... was arbitrary—was just one truth among many possible truths. I could see no reason to think," he would later write, "that humankind was the only example of civilization, unique in all the universe. I imagined there could well be other forms of intelligent life elsewhere" (Drake & Sobel, 1992, p. 5). He also realized early on that these ideas could sound a little crazy to his conventional, middle-class family, neighbors, and schoolmates. So he didn't talk about them much.

Drake attended Cornell University in Ithaca, New York, where he planned to study airplane design, which had sounded fascinating when he chose which college to attend. But the reality turned out to be nothing like what he had imagined. He discovered another field, however, that looked promising: electronics. The major, known as "engineering physics," was demanding—the most difficult in the school. However, Drake easily scored 100 percent in the school's notorious winnowing course in basic electronics (no one else had ever done that). He also enrolled in an elementary astronomy course. There, he thrilled to the views he saw through his professor's 15-inch telescope—Jupiter with its stunning, colorful beauty, and its four giant moons, first discovered by Galileo. For the first time Drake saw for himself that these were other worlds, three-dimensional spheres.

"At that moment," he wrote, "I was smitten. Literally star struck. I felt the same kind of thrill I imagine that Galileo himself must have experienced" (Drake & Sobel, 1992, p. 8). It was an important moment. By summer, Drake was building his own 6-inch telescope in his family's basement, and grinding his own mirror.

As Drake pursued other astronomy courses, though, no one ever mentioned the question of extraterrestrial life. So Drake picked up the hint this subject was forbidden and kept quiet about his long-held thoughts on the subject. Then Otto Struve came to campus in 1951 as the invited speaker for a prestigious series of talks known as the Messenger Lectures.

Otto Struve was the founder of McDonald Observatory, a former director of the Yerkes-McDonald observatories, and currently professor of astrophysics at UC Berkeley. Descended from German-Russian aristocracy, Struve looked, sounded, and acted the part of a gentleman, a member of a four-generation dynasty of noted astronomers. Considered the father of astrophysics, he was known for his bold, fresh theories, well buttressed by substantiating evidence. Struve was an expert in the evolution and structure of stars, knowledge derived through that basic tool in the astronomer's tool kit: the spectrograph. Astronomers use the spectrograph to break up light from a star into its components, shown in the form of spectral lines or spectra, which provide a sort of bar code of information, captured as a photograph.

Drake attended all three of Struve's Messenger Lectures, which centered on methods for squeezing more and more information from stellar spectra. Struve knew how to milk every ounce of information from a star's spectral lines, and in these lectures he outlined some of his latest successes. Drake was fascinated by all three lectures, but the third was the most startling. In this lecture, Struve discussed his observations of stellar rotations. He found he could deduce the speed of a star's spin from the width of its spectral lines, as the bar code becomes spread out by the rotation. When a star is spinning, it is both coming and going, so it is both blue-shifted and red-shifted. (Red-shifting and blue-shifting are visual Doppler effects that tell astronomers what is moving away from us and what is moving toward us. Like the rising pitch of a train whistle as it approaches and the falling pitch as it recedes, a stellar object that is moving away from us appears redder than it ordinarily is and one that is moving toward us looks bluer.) For Struve, reading the rate of change from red-shift to blue-shift and back again was like having a speedometer for stars. When he studied the rotational speed of stars he found some startling results, given that stars come in all sizes—from many times the size of our Sun (which is an average-sized star) to much smaller and fainter than our Sun. The surprising news was Struve found all the huge, massive hot stars were whirling really *fast*. Smaller stars like our Sun were spinning much slower. There was no middle ground.

Struve concluded the fast-spinning stars were lone stars. But those that moved slower, he hypothesized, were impeded in their spinning by an invisible companion star or planet. So, he concluded, stars the size of our Sun may commonly have a hidden companion—very likely one or more planets. Listening to Struve's account, Drake was galvanized. The first basic necessity for the existence of life on other worlds just became much more likely: There probably *were* other worlds beyond our solar system—lots of them! Today, astronomers have detected many planets orbiting other stars, confirming Struve's conjecture. But in 1951, this was big news. Struve had just increased the possible number of planets from the nine in our solar system to something in the neighborhood of 99 billion. That was not all. He also said that, based on the number of planets likely existing in the universe, one could reasonably deduce life could exist in many other parts of the cosmos.

Drake was amazed. No longer alone in his thoughts, he had gained an important ally. A much-respected astrophysicist had just announced to a room full of scientists that life might exist on other worlds, in other solar systems, circling other stars.

In 1952 Drake completed his degree at Cornell with honors in engineering physics, followed by three years serving in the Navy, where he put his knowledge of electronics to work. At the end of his military service, in 1955, he entered the Harvard University graduate program in astronomy. By chance, he obtained a summer position working for the radio astronomy laboratory—with his Cornell degree and his Navy experience, he was uniquely well prepared in this field. And so Frank Drake became a radio astronomer. From that moment everything flowed logically to a Harvard Ph.D. in radio astronomy. Suddenly Drake found himself building a career on the cutting edge of a brand new science.

Drake compares radio astronomy to using a different octave on a piano keyboard. Visible light is the stuff of optical astronomy, providing the sort of information Galileo and Otto Struve studied. But radio astronomy "listens"; it collects data from a different part of the electromagnetic spectrum—from a different "octave." In addition to light waves and radio waves, stars and other objects in space give off other forms of radiation, other "octaves," including gamma rays, X-rays, and infrared waves. Most of this information, though, cannot be collected by ground instruments working beneath Earth's atmosphere, which screens out these rays. X-ray and gamma-ray astronomy have to be done from space-borne satellites orbiting high above Earth's atmosphere. So, in 1955, these types of astronomy did not yet exist. Radio astronomy, however, could work very well from the ground. (Another advantage of radio astronomy is its "perpetual night," moreover. Since light does not affect radio signals, a radio telescope can work around the clock—a nice advantage.)

To gather radio waves from space, radio astronomers use enormous curved dishes that look like giant satellite dishes. These huge antennas can be used in interesting ways by astronomers. In a technique similar to radar mapping, astronomers can bounce radio signals off a planet's surface in a careful, systematic pattern, and by accurately measuring the time taken for the round trip, they can produce a reasonably accurate topographical map of the surface, even one hidden beneath a thick cloud cover like Venus's.

Radio telescopes are also used for imaging radio emissions from other objects in the cosmos. Because the radio telescope is looking at a different kind of emission in the sky, everything in radio astronomy tends to look completely different from visible-light images. Objects that give off bright, luminous optical light generally are not strong radio sources. (In fact, even relatively strong radio emissions from space are usually faint—that's why such large antennas are needed.) So if you looked at a radio picture of the cosmos, you would see no stars and the Moon would barely be visible.

The discovery of radio astronomy in the 1930s and the early 1940s opened up new ways of "seeing" objects that were otherwise so dim that their light was absorbed by clouds of dust. Radio waves traveled through the dust, they were not blocked by it, in much the same way you might clearly hear the melody of a tenor saxophone playing in a smoke-filled room, but might not be able to see the bartender clearly. Dutch astronomer Jan Oort was one of the first scientists to recognize what a valuable tool a radio telescope could be, especially in exploring the structure of our own Galaxy. He found, while large clouds of dust blocked the view looking toward our Galaxy's center, radio waves easily penetrated the dust and could be picked up by a radio telescope. It was not the same as an optical view, true, but information about the structure of objects in the center of the Galaxy could be deduced from radio emissions, and Oort found he could perceive radio emissions even from the other side of the Galaxy. Oort figured out how he could use this information to study the rotation of the Galaxy and estimate distances to clouds of gas. He could then use this information to map how matter was distributed in the Galaxy.

A student of Oort's, H. C. Van de Hulst, began figuring out the spectra typical of elements, in much the same way this information had been tested and recorded for visible light. Starting with the most abundant element in the universe, hydrogen, he quickly discovered hydrogen had a clear spectral line at 1420 MHz. Another way to designate this spectral line would be by its wavelength, or frequency, which measures 21 centimeters (8.27 inches). From the progress made by Oort and Van de Hulst, radio astronomy slowly began to grow.

By the time Frank Drake completed his degree at Harvard in 1958, the National Radio Astronomy Observatory (NRAO) had just been established in Green Bank, West Virginia, and Drake immediately received a job offer. (Coincidentally, Otto Struve would become the first full-time director there just two years later.)

Along the way, Drake had begun thinking further about communicating with extraterrestrial civilizations. How could astronomers reach out to beings on other worlds? What "postal system" would they use? It seemed logical that they, too, would realize they were not alone in the universe. It came to Drake that a radio telescope would be a good receiver for alien messages. By the time he arrived at Green Bank, he had made many calculations to weigh how well one instrument or another might do the job. As a guideline, he assumed alien equipment might be no more powerful than the best we had on Earth. The abiding question was always: How close would they have to be for communication to take place? Drake had begun his work in radio astronomy with a small, 25-foot telescope at Harvard. He made his calculations: That telescope was too small. The alien source would have to be closer than the nearest star. Big dishes were required for more distant sources because radio signals emitted from space were so faint. The 60-foot telescope at Harvard was also too small. But the 85-foot telescope they were building at Green Bank was different. It could be big enough—barely big enough, but big enough. So Drake had his eyes on the 85-foot telescope from the beginning—not only because of its size, but also because it would use a new, better technology than the others he had worked with.

In early 1959 he received encouragement from the current director at Green Bank, Lloyd Berkner, and Drake began work on the project that would establish his fame: Project Ozma (named after Princess Ozma in the

Oz books—*The Wonderful Wizard of Oz, Ozma of Oz,* and so on—by L. Frank Baum). Ozma would search for radio signals that might be a message from an intelligent extraterrestrial civilization. The name was exotic and romantic, evoking the idea of alien lands with strange, unknown creatures, but Drake designed the project economically and logically, building the receiver to serve several other observational goals after Ozma was finished. Ozma was also to be a short project—only two weeks long—squeezed in among the other projects of the NRAO.

The plan was to conduct a small survey. If a signal came in, it would be enormous news. Otherwise, Drake had no plans to announce the experiment before, during, or afterward to avoid media hoopla and embarrassment. Then something happened that changed those plans. In September 1959, a paper was published in *Nature,* written by two physicists at Cornell, Giuseppe Cocconi and Philip Morrison. For the first time in print, these scientists pointed out that existing radio telescopes had sufficient sensitivity to pick up radio signals from far-off stars. They granted that a thorough search would take time and persistence, but they pointed out if no one ever searched, nothing would ever be found. The newspapers were soon buzzing. A search for extraterrestrial intelligence suggested by well-known, respected scientists. This was news.

By this time, Otto Struve had become director of the NRAO, and he was enraged at being scooped. Ozma had been planned for nearly a year. Now someone else would get the credit for the idea. He decided to announce the "secret" project when he delivered a series of lectures at MIT a few weeks later.

Drake, meanwhile, was under pressure to run the survey. Cocconi and Morrison had suggested the same frequency that Drake already planned to use—the 21-centimeter line—and their article helped validate his choice. Drake had chosen the 21-centimeter line for Project Ozma because it fit well with another project that was already running on the same telescope. However, Cocconi and Morrison had a better reason for using that wavelength to listen for communications from intelligent civilizations: It is the wavelength of hydrogen radio emissions, and because hydrogen is the most abundant element in the universe, this would be common knowledge for any advanced civilization in the universe, and thus the natural listening and broadcasting frequency. Moreover, the emission activity is not extensive in that neighborhood. And, as a topper, nearby, the spectral emission line for the hydroxyl OH (oxygen and hydrogen) at 18 centimeters creates a sort of "water hole," an intellectual pun implying water, essential to all life as we know it, formed of two hydrogen atoms and one oxygen ($H + OH = H_2O$).

The *Nature* article pointed out that the long-wavelength region near the hydrogen line was ideal because it represented not only the most abundant element in the universe but also the most fundamental (one proton, one electron). Also, Drake realized, he would run into very little radio interference in this region of the electromagnetic spectrum.

Ozma could not sweep the skies for a full-sky search; it would observe just two stars, Tau Ceti and Epsilon Eridani. Near the beginning of the search, one moment of great excitement occurred when the big telescope picked up a signal, but the signal turned out to have an Earthly source. The rest of the time played out with no results. But it was a beginning. No one had really expected to find an advanced civilization on the first try. All in all, the search cost no more than $2,000 to do.

FIRST CONFERENCE ON SETI

About a year later, a plan for the first conference regarding the search for extraterrestrial intelligence (SETI) began to take shape. J. Peter Pearman, a staff officer of the Space Science Board of the National Academy of Sciences, contacted Drake about the idea. Pearman hoped to encourage government interest in researching the possibility of life on other worlds. Immediately the two began putting together a list of invitees, with plans for a conference at Green Bank.

Giuseppe Cocconi and Philip Morrison, authors of the seminal paper in *Nature,* headed the list of invitees. Drake suggested Dana Atchley (knowledgeable about ham radio and electronics), who had donated the parametric amplifier Drake used in Project Ozma. Barney Oliver, an inventor with Hewlett Packard who had visited Project Ozma, was high on the list. And Otto Struve, as well as Su Shu Huang, a Chinese-American astrophysicist with NASA who had worked as a student with Struve investigating the kinds of stars that might support habitable planets.

Both Pearman and Drake came up with the name Carl Sagan, who had contacted Drake about some research the radio astronomer had done with the 85-foot telescope on the temperature on Venus. A continuing correspondence had brought Drake's attention to Sagan's interest in exobiology. Also, Pearman knew about Sagan's position on the Space Science Board Committee on Exobiology and on the NAS Panel on Extraterrestrial Life. At 27, he was the youngest in the group of invitees to the conference. Drake described him as "dark, brash, and brilliant" (Drake & Sobel, 1992, p. 54), and Drake would later write of Sagan: "He knew more about biology than any astronomer I'd ever met, and was fast making a never-before-heard name for himself as an 'exobiologist'" (Drake & Sobel,

1992, p. 47). Two other names linked with Sagan were also on the list: Joshua Lederberg and Melvin Calvin. Both had an abiding interest in what Lederberg had dubbed exobiology and both were Nobel laureates—actually, Calvin received the news of his award while at the Green Bank conference, prompting a jubilant flow of champagne.

Drake jokingly remarked that with invitees from the fields of astrophysics, astronomy, electronics, and exobiology, all they needed was someone "who's actually spoken to an extraterrestrial." Pearman had someone in mind for that, too, he claimed: John C. Lilly. A physician and researcher in medical physics and experimental neurology, Lilly had pursued experimental research in consciousness and communication. He came about as close as anyone could to attempting communication with an extraterrestrial, because he was exploring the possibility of communicating with dolphins, which were then considered to have an intelligence equal to or greater than humans. The prospect of communicating with nonhuman intelligence was intriguing. Over the following years, Lilly never succeeded in showing he *could* communicate with dolphins, and his dolphin research eventually fell into disrepute, due in part to his lack of experimental controls and inadequate record keeping. His interest shifted to other aspects of consciousness, including sensory deprivation. However, at the first SETI conference, his accounts of his experiences with dolphins charmed his audience (as, in fact, they would continue to do for other audiences for several years). His contributions were appreciated on a par with the other invitees, nearly all of whom attended. Only Cocconi was unable to be there.

THE DRAKE EQUATION

Just in time for the first SETI conference, Frank Drake devised a method for thinking about how to conduct a search for extraterrestrial intelligence. Known as the "Drake Equation," it isn't really an equation. So many questions surround the search for intelligent civilizations that Drake devised his "equation" to focus on them one by one. He wanted a way to estimate how many intelligent, communicating civilizations might exist in our Galaxy, and his equation is more of a way of thinking about these questions than a specific mathematical formula. It looks like this:

$$N = R\, f_p\, n_e\, f_l\, f_i\, f_c\, L$$

This shorthand equation basically asks questions to determine how many intelligent, communicating civilizations might exist in our Galaxy, the Milky Way (*N*). Drake figured that by calculating the product of several factors, one could come up with a reasonable estimate. That is: **N,** the **N**umber of communicating civilizations, can be estimated by multiplying together the answers to these basic questions:

- **R** What is the ***R**ate of star formation* in the Galaxy?
- **f_p** What is the ***f**raction* of those stars that have ***p**lanets*?
- **n_e** What is the ***n**umber* of ***e**nvironmentally* appropriate planets orbiting each of those stars?
- **f_l** What is the ***f**raction* of those planets that evolve ***l**ife*?
- **f_i** What is the ***f**raction* of those planets that evolve ***i**ntelligent* life?
- **f_c** What is the ***f**raction* among the intelligent-life-bearing planets that can send ***c**ommunication*?
- **L** What is the ***L**ength* of time (in terms of a fraction of planet's life) that a communicating civilization may survive?

The big problem is that no one knows exactly what numbers to use in the equation—because no one really knows the answers to all the questions Drake posed. Still, Drake staked out a place to begin.

HOW MANY ARE OUT THERE?

Drake began by thinking the development of life requires a planetary system, something like our solar system, with one or more planets that might be home to living organisms. Beginning with that idea, each element in the Drake Equation stands for one of the several factors influencing how many communicating civilizations could exist in our Galaxy.

At the conference, Sagan and the rest of the attendees tried coming up with reasonable estimations for all the factors in Drake's equation to see if the idea of trying to communicate sounded plausible. For example, by plugging in 200,000,000,000 (two hundred billion) for the rate of star formation in the Galaxy, they would have a place to start. The rest of the numbers might be:

- **R** What is the ***R**ate* of star formation in the Galaxy? (200,000,000,000)
- **f_p** What is the ***f**raction* of those stars that have ***p**lanets*? (.20)
- **n_e** What is the ***n**umber* of ***e**nvironmentally* appropriate planets orbiting each of those stars? (3)
- **f_l** What is the ***f**raction* of those planets that evolve ***l**ife*? (.50)
- **f_i** What is the ***f**raction* of those planets that evolve ***i**ntelligent* life? (.20)

- f_c What is the *fraction* among the intelligent-life-bearing planets that can send *communication*? (.20)
- ***L*** What is the ***L****ength* of time (in terms of a fraction of planet's life) that a communicating civilization may survive? (1/1,000,000)

At the meeting, they discussed each factor, each question, one by one. There was room for a great many differences of opinion. And variations in the answers could wind up with widely varying results ranging from one to millions. Of course, all the numbers they plugged into the equation were very tentative, very rough estimates—and they still are rough today. However, current estimates for the number of stars in our Galaxy are at least 200 billion. Of course, many other galaxies besides the Milky Way exist—100 billion galaxies or more. Each one of those galaxies may also contain at least 200–300 billion stars, each of which may be a "home star" to other intelligent civilizations.

In 1961, only one planetary system was known—ours. Everything else was speculation. (Since then, astronomers have detected many extrasolar planets, and are constantly discovering more. A much more accurate estimate may now not be far off. Currently, some researchers think nearly all the stars in the Milky Way may have planetary systems!)

Most of the other numbers were even more controversial than these. And when they changed the estimates, the outcome changed.

And another question: What percent of those planets that *could* support life actually do host living organisms? Some scientists think "life is easy." That is, the conditions that allowed life to begin on Earth are probably not so rare and, anyway, life may not require those exact conditions. Life on Earth exists in many extreme conditions. Could valid estimates for this factor be as high as 100 (1.0)?

On the other hand, some of the most important elements required for life on Earth are actually very rare. Carbon, nitrogen, oxygen, sulfur, phosphorus, calcium, and iron are all among the scarcest elements in the universe, produced only in supernova explosions. So, other researchers are not nearly so optimistic, maintaining that Earth's circumstances were and are rare. They might put the estimate nearer to 0.

Among those planets where life begins, what fraction develop intelligent life? Again, the range of estimates is broad. No one really knows. Some researchers think that intelligence is such a clear advantage that intelligent life will always develop if it has a chance. Others point to the ancient cockroach. What intelligence does this lowly insect have? Yet, it thrives today—almost completely unchanged since it first appeared on

Earth more than 320 million years ago. Apparently a lack of literature and mathematics has had no real negative effect on this insect's survival. So, again, estimates range from nearly 100 percent (1.0) to just above 0.

What about communicating? No one really knows how unique our communication capabilities are. Our abilities are by far the most advanced on Earth—including the means to broadcast our communications using radio waves. How many intelligent civilizations would develop these skills? Since this is basically an unknown quantity, most researchers settle for 50 percent (0.5).

By the time you get to the last factor, you have probably narrowed down the 200 billion original candidates a lot. One last factor remains in Drake's list: How long does a planet survive, after a communicating civilization has developed there? No one knows. Our own civilization has only been "communicating" in the cosmic sense since the birth of radio and TV, about fifty years ago! Scientists calculate that Earth and Sun will probably last a total of about 10 billion years. However, what portion of that time will our civilization survive? Let's be optimistic: If we survive another ten thousand years, the estimate for the last factor would be 1/1,000,000th.

As for the best frequency for listening, the scientists gathered at the first SETI conference agreed on the 21-centimeter standard already beginning to emerge. They also liked the idea of gathering and communicating at the water hole like animals at a desert oasis. For SETI, therefore, the 21-centimeter frequency. seemed appropriate for several reasons and became the standard for both listening for messages and sending them.

After the meeting, Drake put a sign outside his door: Is there intelligent life on Earth?

It was a joke, but it pointed out the level of uncertainty about the quest, enthusiastic as Drake was. He would later write about that period: "Whimsical and tentative, still, were other people's attitudes toward the subject. One could now talk about the idea aloud, thanks to Project Ozma and the Cocconi-Morrison paper in *Nature*—but not without looking over one's shoulder to see who might be laughing" (Drake & Sobel, 1992, p. 46).

Today, Frank Drake heads the SETI Institute in Mountain View, California, which was founded in 1984.

ARE WE ALONE?

Weaving through the entire length of Carl Sagan's life was the haunting question "Are we alone?" Are there other worlds on which life exists? Even more poignant, are there other living beings like us elsewhere in the universe? Or are the vast reaches of space beyond the protective shell of Earth's atmosphere completely uninhabited? Are all those beautiful expanses of fire and ice an endless void, unobserved by any other beings? Without worlds or civilizations or colonies or cities of sentient, thinking beings?

Nearly every aspect of humankind's entrance into space has begun with the kind of youthful enthusiasm that Frank Drake and Carl Sagan gave to SETI from its earliest inception. Rocketry began with clubs of boys and young men in Germany, the United States, and elsewhere, eagerly trying to send rudimentary rockets streaming skyward. Wernher von Braun and many of his rocket team originally built rockets for the German military (and in the process got themselves involved in a very ugly business), but dreamed of making space travel possible. Today's NASA teams, which have put rovers on Mars and send robotic emissaries to comets and asteroids, begin their projects with great élan and enthusiasm and, when successful, end them with hearty cheers and hugs all around. Success in space has always required both ebullient optimism and hard work.

SETI, of course, has nothing to do with "going there," unlike rockets and spacecraft or satellites. The technology used for SETI would have come into use without the space program. Yet, SETI is the equivalent of seeking signs of civilization in a thick jungle, or expecting a letter from a far-off land. It is the search for hoped-for communication. How reasonable is that hope? To some people, SETI somehow seems frivolous. A little too far out. Yet, from Frank Drake's Project Ozma to such recent searches as the SETI Institute's Project Phoenix, SETI is serious science.

On the other hand, UFO believers often asked Sagan how he could disbelieve in UFOs and alien abductions and yet support the search for extraterrestrial intelligence. Sagan would acknowledge he had a deep personal interest in extraterrestrial life, especially in the possibility of encountering some evidence of intelligent life. But he would also point to the safeguards one must set up against fooling oneself. One must require rigor and especially clear evidence. As he said near the end of his life in an interview on the PBS program *NOVA* (1996):

> [The discovery of extraterrestrial life] would be an absolutely transforming event in human history. But, the stakes are so high on whether it's true or false, that we must demand the more rigorous standards of evidence. Precisely because it's so exciting. That's the circumstance in which our hopes may dominate our skeptical scrutiny of the data. So, we have to be very careful. There have been a few instances in the [past] we

> thought we found something, and it always turned out to be explicable...
>
> So, a kind of skepticism is routinely applied to the radio search for extraterrestrial intelligence by its most fervent proponents. I do not see [in] the alien abduction situation a similar rigorous application of scientific skepticism by its proponents. Instead, I see enormous acceptance at face value, and leading the witness and all sorts of suggestions....

"It's a wonderful prospect," Sagan concludes, "but requires the most severe and rigorous standards of evidence."

Chapter 5

EXPLORING THE PLANETS

Sagan's three years on the West Coast came at an exciting time. He became one of the select scientists to help shape a watershed moment in human history: the birth of the space age. After millennia of wishful thinking, at last humans had thrown off the shackles of Earth's gravity. Rockets soared above the atmosphere. Satellites made their lofty orbits, miles above the surface, as if suspended by invisible wires. Robotic spacecraft—both U.S. and Soviet—had begun the first exploratory missions to the Moon. All this was brand new and stunning. *Sputnik*'s first lonely orbit had changed forever the way humans would perceive their place in the universe.

For both Americans and Soviets, the time also carried with it a sense of frenzy—an all-consuming competitive tension. And by May 25, 1961, President John F. Kennedy had announced the U.S. intention to initiate a human presence on the Moon within 10 years.

Meanwhile, planetary exploration efforts had already begun. The first Soviet efforts to send probes to the planets in the late 1950s had failed, but the U.S. Mariner program had a shot at being first to succeed, so the push was on. Planetary scientists were delighted at the opportunity to ride the coattails of the momentum created by primarily political motives.

In part because of this hyperactive atmosphere, in part because of his own preexisting ambitions, Sagan spent these West Coast years in a highly stepped-up level of activity. He accepted a furious schedule of telephone and personal interviews, presentations, and lectures. In addition to this, he continued to explore the possibility of life on other worlds, research the origin of life, defend his greenhouse theory about Venus, advise the NAS exobiology group, consult for the RAND Corporation, meet with Lederberg, and experiment with the simulated development of microbial life on Mars. He also agreed to design an instrument for an astronomical balloon and joined the editorial team for an important astronomical journal. To top off the workload, he began to consult as a member of the science team for *Mariner 2*, a flyby mission to Venus (and, as it turned out, the first successful mission to any planet). It was as if, for all his training in astronomy, he had forgotten how many hours there are in a day. Also, in October 1960, Lynn and Carl's second son, Jeremy Ethan, was born. It was a heavy workload for anyone, but Sagan tried to juggle it all.

THE FIRST MARINER MISSIONS

From the first missions planned by NASA—at the encouragement of rocket expert Wernher von Braun—"redundancy" had become a key engineering principle. (And one that NASA continues to favor, whenever possible.) The idea is: Always have a backup, or at least a contingency plan. That is why most important planetary missions have been sent in pairs (a tradition NASA still follows when feasible). The strategy paid off when NASA's Jet Propulsion Laboratory (JPL) launched its first planetary mission, two Mariner spacecraft, *Mariner 1* and *Mariner 2*, were designed to visit Venus. Known as "flyby" missions, these two spacecraft were designed to fly by the planet within about 25,500 miles, nothing fancy by today's standards, but very exciting just the same, because no one had ever seen any of planet from any closer than people can get by building an observatory on one of Earth's high mountaintops, or, perhaps, a telescope-bearing balloon launched to soar high into the stratosphere. As *Mariners 1* and *2* flew by, the two spacecraft would take close-up readings to verify or disprove the highly disputed temperature and atmospheric pressure measurements made from Earth.

However, when *Mariner 1* was launched on July 22, 1962, the launch rocket suddenly veered off course after liftoff. Both rocket and spacecraft had to be destroyed, so *Mariner 1* never reached orbit. All went well, though for *Mariner 2*, launched a month later. So the redundancy strategy had paid off.

For Sagan, being included on the Mariner science team was a priceless opportunity. This would turn out to be the beginning of a lifetime of consulting on planetary missions for NASA's JPL. Located in Pasadena, California, and managed by the California Technical Institute (CalTech), JPL would become the heart of NASA's planetary exploration program—so Sagan, once again, was in exactly the right place at exactly the right time. In fact, after naming off some of the most cherished projects he had engaged in thus far in his career (including planetary exploration, the effort to simulate the steps in the origin of life on this planet, and SETI), he remarked, "Had I been born fifty years earlier, I could have pursued none of these activities.... Had I been born fifty years later, I also could not have been involved in these efforts" (Sagan, 1973, p. xvii)—at least not in the initial stages of them. He was born at a time when he could participate in the first efforts to explore the solar system with spacecraft and a human presence. Similarly, technology and science had just arrived at the stage of sophistication that would allow the beginning efforts to understand life and its components (although, even with the advent of cloning, much still remains to be discovered). He was also among the first group of scientists to attempt serious communication with civilizations on other worlds.

When the two Mariner spacecraft were still in development, Sagan published his article, "The Planet Venus," in the March 24, 1961, issue of *Science*. It was his first major publication, in one of the country's most respected peer-reviewed journals, published by the American Association for the Advancement of Science. Based partly on the Venus portion of his dissertation, this article laid out Sagan's hypothesis that a runaway greenhouse effect on Venus was responsible for the extremely high temperature measurements deduced by researchers using Earth-based instruments. These measurements, though, remained contested by many planetary scientists. Since no more direct measurements were yet available, Sagan risked being proven wrong as soon as *Mariner* flew by Venus and measured the atmosphere up close and personal.

One of the processes of science involves devising a theory that explains observations other scientists have made and then making a testable prediction for evidence not yet gathered. Such a prediction, if proven true, can serve as powerful evidence for a theory. But it can also be a risky business for the theorist. If the prediction turns out to be correct, the scientist is a hero. Otherwise, he or she can easily look foolish. That is exactly the situation set up by Sagan with his *Science* article. When the Mariner spacecraft flew by Venus about nine months later, everyone would know the answer.

Mariner 2 weighed about 447 pounds, with a hexagonal base measuring only about 3.5 feet across and a little more than a foot thick. The total span of its solar panels was about 16.5 feet, and instrumentation mounted above the base raised the spacecraft's height to about 12 feet. During its mission, the spacecraft would fly by Venus and collect information about the planet's atmosphere, magnetic field, charged particle environment, and mass. Racing to reach Venus before any Soviet spacecraft got there, NASA built the spacecraft quickly, in fewer than twelve months. It looked a little like a huge dragonfly, but for an interplanetary spacecraft, it was petite.

The spacecraft itself had to come together rapidly—including communications, data encoding, computing, timing, attitude control, power control, battery, and battery charger, as well as the attitude control gas bottles and the rocket engine. When it was time to add the scientific instruments, the members of the science team had to have the equipment ready to go. If any instruments were not ready, the program manager let it be known he would just strap on a hunk of lead instead. Getting there was NASA's main goal, so the scientists had to plan and build their instruments rapidly. Sagan worked on the infrared radiometer.

During development, Sagan campaigned for including a camera on the spacecraft, to "answer questions we were too stupid to ask" (Davidson, 1999, p. 118). Sagan was visually oriented, and he recognized that most of the public was, too. Perhaps, through some break in the cloud cover, we would see some clues to the history of the planet, some evidence of life there in the past or present, or maybe just gain the satisfaction of seeing the dense atmosphere of our nearest neighbor up close. However, most scientists tend to prefer to see their vistas in numeric terms, and the case for the cameras was lost.

Still, Sagan did not need a camera for the answer that was so key to the success of his career at this point. Much of his stature as a scientist rested on the stand he had taken on the existence of a greenhouse effect on Venus. One instrument onboard, a microwave radiometer, could do him in if it reported the presence of a phenomenon known as "limb brightening." He had based his theory on the report from the rooftop scientists at the Naval Research Lab (NRL) and their high microwave reading, which he and they had interpreted as a reading of surface radiation, indicating very high surface temperatures. Not every scientist agreed with this interpretation, though. Some researchers maintained the NRL measurement might just indicate a strongly ionized region high in the atmosphere, while the surface might be relatively cool. If so, as *Mariner* passed Venus, the microwave detector would pick up an intense signal at the limb, or edge of the disc. An intense reading in this location is known as limb brightening. In that case, Sagan's greenhouse theory and his vision of the atmosphere of Venus would lose its persuasiveness. If, however, the microwave signal grew weaker at the limb, that is, if the detector discovered limb darkening, Sagan's greenhouse theory could still be in the running (though not proven).

Had Sagan offered a viable explanation of the facts? Or had he missed? Relentless time would tell, and, like a gunfighter of the Old West or a card player at a poker table, Sagan consciously chose not to blink.

PLAYING THE HAND

As Sagan considered the strengths—and weaknesses—of his hand, not only did his greenhouse theory attract Frank Drake's attention, but now, publication in *Science* also garnered interest from perhaps the most prestigious school in the country: Harvard University. Fred Whipple, known for his prowess in the realm of the solar system, and Donald Menzel, a specialist in solar studies, and both top astronomers at Harvard University (Whipple headed the Astronomy Department), contacted Sagan about leading a colloquium on the Cambridge, Massachusetts, campus. Even though Whipple had championed the ocean-covered-surface vision of Venus, the two invited Sagan to Harvard to share his views at a colloquium on planetary studies. The colloquium was a great success (Sagan was nothing if not a natural orator and communicator). It went so well that Whipple and Menzel persuaded the school to offer him a position as lecturer on the astronomy faculty and a joint post at the Smithsonian Astrophysical Observatory. It was, by any standards, a savory offer.

Sagan played it cool, though. He said he could not accept a position as lecturer; they would need to upgrade their offer at least to a position as assistant professor (the next rung on the ladder of academic ranking). The move showed amazing confidence—most recent Ph.D. recipients would gladly accept anything Harvard had to offer. Surprisingly, though, Harvard agreed on the bargain. (Whipple and Menzel had supported Sagan in glowing terms, touting him as extraordinary.) Carl Sagan, it seemed, had arrived, and his future was secure. He negotiated to wait a year before taking his position at Harvard so he could conclude the projects he had in progress in California.

LIGHTING UP FIREFLIES

Sagan's Miller Fellowship had about run out. However, Lederberg had used his considerable influence to secure a short-term appointment for Sagan as assistant professor of genetics at the Stanford University School of Medicine (a sizable honor, since Sagan's doctorate was in astronomy and astrophysics). Sagan joined with Cyril Ponnamperuma in some key origin-of-life experiments Ponnamperuma was conducting at NASA Ames Research Center (ARC) in nearby Mountain View, California, on the San Francisco Peninsula. Like Sagan, Ponnamperuma was fresh from Melvin Calvin's lab at UC Berkeley, where he had taken part in some of Calvin's experiments on the origins of life. Ponnamperuma now had a position in the new Exobiology Division at Ames.

Following up on the Miller/Urey experiments, Sagan and Ponnamperuma planned an attempt to synthesize—under the conditions believed to be prevalent during Earth's early years—the molecule adenosine triphosphate (ATP).

ATP can be thought of as a sort of organic battery, a molecule present in all living things, providing the main source of usable energy at the cellular level. The adenosine part of the molecule is composed of ribose, a sugar having five carbon atoms, and adenine (one of the four fundamental bases of DNA), which contains nitrogen. This compound is linked to three phosphates, each composed of one phosphorous atom and four oxygen atoms, connected through bonds that easily break, producing energy. This is the power that produces the flash in fireflies, as well as other activities of living cells. Successful production of this molecule would be a persuasive next step.

Experimenting in his dungeon lab at the University of Chicago, Miller used simulated lightning to jolt organic molecules into existence. Now, in the role of theorist for Ponnamperuma's team, Sagan postulated that extreme ultraviolet (UV) radiation from sunlight in the early Earth environment might have induced the process. Ponnamperuma directed the laboratory end of the collaboration, and lab technician Ruth Mariner did most of the hands-on work. The process involved taking a mixture of gases including phosphorus (together representing Earth's early atmosphere in this scenario) to Berkeley's cyclotron lab, where they exposed the gases to radiation (representing sunlight falling on the atmosphere). They mixed one of the products, adenine, with water, phosphorus, and a sugar (which might have come together in, say, a puddle or small body of water on Earth's surface). And they exposed this mixture to UV radiation (again simulating the harsh sunlight that fell on early Earth). They hoped to see formation of organic products, including ATP.

According to the chromatography tests Mariner used to identify the products of the experiment, they had produced a large amount of adenine and adenosine, plus a small amount of ATP. To be sure of their finding, they ordered a batch of ground up firefly tails from a bioresearch supply house. When they exposed the dead firefly tails to the ATP, sure enough, the ground up tails lit up! The team published their results in the July 20, 1963, issue of *Nature*, a much-respected British general-science journal, first published in 1869.

The published results created a stir of interest in the scientific community and led to speculation that ATP formed in this or some similar way might have provided all the energy living cells needed prior to the time photosynthesis began to take place. Sagan conjectured a few years later that ATP may have formed nonbiologically as a free molecule and then rained down on living organisms, providing them with free energy. And that would explain why ATP was a universal energy source for all living organisms. At the time, many scientists, including Sagan, thought science was very close to proving how life began, not only on Earth, but also, he speculated, on Mars, some billions of years ago. Today, more than forty years later, we still have not resolved this question, and scientists still have not confirmed any past existence of life on Mars, despite a growing body of evidence that the conditions must once have been right for it.

In the case of the Ames ATP experiments, Stanley Miller discounted them because of what he considered a faulty premise: What evidence is there that such large quantities of phosphorous were at hand on early Earth for the series of reactions outlined in the experiment's scenario? Also, after publication of the *Nature* article, difficulties arose in repeating the experiment. (A repeatable experiment is a key ingredient for the development of a viable hypothesis.) Some attempts were made to redo the experiment, but in the end Ponnamperuma simply stuck by his original results, garnering some criticism for sloppiness in his laboratory technique and a lack of concern for reporting negative results as well as positive ones. Instead, he simply moved on. Sagan's part in the teamwork was the theoretical basis, which he had outlined cleanly and clearly. Not all the projects he had going during this period went even that well, however, and the reasons were complex.

OVERCOMMITTED AND UNDERCOMMITTED

One of those projects was an experiment Sagan wanted to launch on an astronomical balloon known as *Stratoscope II*. Its predecessor, *Stratoscope*, had been used by a Princeton University team since 1957 to carry a robotic telescope high into the stratosphere. From there, the telescope could make observations with less interference from water vapor and atmospheric turbulence. As the group was developing *Stratoscope II*, Sagan flew back east to New Jersey to present an idea he had for observing Mars with the new balloon's robotic telescope. He persuaded the team leader, Martin Schwarzchild, that an infrared spectrograph would be a highly useful addition to the instruments already used, possibly enabling the identification of living microbes on Mars. The prospect would also be attractive to NASA because it could enable the United States to scoop the Russians by discovering Martian life first. Also, the plan could bring fame to Princeton University. The team liked the idea and adopted it. Sagan promised to develop the specialized infrared detector that would be needed.

Sagan and Frank Drake traveled to Dallas, Texas, to visit the Texas Instruments lab of Frank Low, who was developing an advanced infrared detector that used germanium. Sagan seemed enthusiastic. But, for some reason, he never did get the job done. It seemed like a good idea, though, and other UC Berkeley scientists completed the infrared spectrograph and the instrument flew aboard *Stratoscope II* in 1963, obtaining a spectrum of Mars at 78,000 feet. However, it found no signs of life.

The incident illustrates some of Sagan's strengths and weaknesses. He had vision, came up with practical ideas, could make connections, and could persuade people to commit to a project. But then he did not necessarily follow through on the details himself. Sagan later wrote to Gerard Kuiper that he had overcommitted himself during the Berkeley years and had not followed through as he felt he should have.

Meanwhile, on another front, from his wife Lynn's point of view, he had drastically undercommitted to his family. In 1962, when Dorion was three and a half years old and Jeremy was two, Lynn told Carl she and the boys were leaving him, and she was filing for divorce. Carl was stunned. He took her for a long walk and tried to talk her out of her decision. He reminded her how well his career was going (not realizing that his overcommitment to his career was a large part of the problem). Lynn had a long list of deep-seated complaints: He was never home. He didn't help with running the household, raising the children, or handling the finances. She was exhausted and had no time for her own work on her Ph.D.

Later, Lynn found a note that Carl had written to himself during this time. It was headed, "Lynn's present difficulties," with an arrow and these added words: "And mine, and mine." He berated himself for belittling her and for his inattention. He listed things he thought had gone wrong and plans for setting them right. He also expressed self-criticism for not telling Lynn how much he loved her.

However, the cause of strains and rifts in a relationship is never one-sided, and in some ways Lynn undermined their relationship herself and was frequently unsupportive to Carl, as she has since been the first to admit (and did so to Carl in later years). But there was no setting it right. Even if he had been more attentive, they had embarked on a steep and winding path at best—a two-career family with children—a path that required great consideration, cooperation, and compromise on both sides.

For both partners, careers were a high value, and had it not been so the world would have been cheated out of one or both fine scientists. Also, Sagan's mother had not raised Carl with consideration, cooperation, or compromise in mind. She placed him at the center of everything and established that expectation in his mind. This sense that he was entitled to be the focus of attention contributed to Sagan's enormous, winning confidence. But for Lynn, it created an expectation that did not leave enough room for her to fulfill her own expectations of herself.

From Lynn's side, finding that Carl did not respond well to criticism, she quickly developed a pattern of making no argument (but going away angry) or waiting until they were at a public meeting to critique his ideas (which was embarrassing to him). Sagan did not yet have a sense that synergy and partnership can have beautiful results. That would take the maturing influence of time and a different partner.

Lynn moved out and moved in with Carl's cousins Arlene and Eugene Sagan, who lived nearby. Her career was in limbo, with a Ph.D. at Berkeley not yet finished. She had been under pressure from Carl to finish up quickly because he needed to leave for Cambridge and his position at Harvard. Their children were distraught over the separation, and Lynn worried about them as she moved them into a small apartment and looked for ways to make ends meet.

Sagan, meanwhile, was not doing much better. He would later express dismay that he let much of the work on *Mariner 2* slip through his fingers. It was unbelievable, the boy who longed to visit the planets had the opportunity to work on the first interplanetary spacecraft and he blew it. He was just too torn up inside to pay attention.

MOVING BACK EAST

Before embarking on his career at Harvard, Sagan made one last entreaty to Lynn to patch up their fractured relationship. Lynn was appreciative, gentle, but firm. She wrote to him, "Carl, you are a beautiful person. I have watched you grow and I feel you have the potentiality for true and mature love. Just not with me. You know that" (Poundstone, 1999, p. 70).

Once relocated in Cambridge, Sagan made yet another overture to Lynn to come back to live with him. For pragmatic reasons, she finally accepted the invitation—as a roommate. The East Coast offered better opportunities for her in her field. So she moved east with the kids and, together with Carl, took up residence in Cambridge. Not long afterward, though, she moved out, finalized the divorce in 1963, and moved on.

Lynn went on to complete her Ph.D. in genetics at UC Berkeley in 1965. During this period she also developed materials for teaching science. From 1967 to 1980 she was married to crystallographer Thomas Margulis, and she is now known as Lynn Margulis. She served on the faculty of Boston University from 1966 to 1988, and during that time she wrote *The Origin of Eukaryotic Cells*, in which she proposed that eukaryotes (nucleated cells) came about through symbiotic relationships between various bacteria rather than through natural selection, a point of view that created tremendous controversy at the time. Since then, however, her views have gained wider acceptance. Yale ecologist G. E. Hutchinson calls this change a "quiet revolution in microbiological thought," for which he primarily credits Lynn Margulis for her "insight and enthusiasm," adding, "Hers is one of the most constructively speculative minds, immensely learned, highly imaginative and occasionally a little naughty" (Margulis, 1999). In 1988, Margulis left Boston for a full professorship at the University of Massachusetts, where she now holds the title of Distinguished University Professor in the Department of Geosciences. She writes prolifically, counting among her many books: *Origins of Life* (two volumes, 1970–71), *Origins of Sex: Three Billion Years of Genetic Recombination* (coauthor with Dorion Sagan, 1986), *Mystery Dance: On the Evolution of Human Sexuality* (with Dorion Sagan, 1991), *What Is Life?* (with Dorion Sagan, 1995), *Five Kingdoms: An Illustrated Guide to the Phyla of Life on Earth* (coauthor, with K. V. Schwartz, 3rd edition, 1998), *Microcosmos: Four Billion Years of Evolution from Our Microbial Ancestors* (with Dorion Sagan, 1986), *Symbiotic Planet: A New Look at Evolution* (1998), and *Acquiring Genomes: A Theory of the Origins of Species* (2002). Margulis was elected to the National Academy of Sciences in 1983, and in 1999 she received the Presidential Medal of Science, awarded by President William J. Clinton.

SAGAN AS MENTOR: THE EARLY YEARS

With his move to Harvard in February 1963, Sagan would begin to gather a group of graduate students, which included a trio that would become solid planetary astronomers: Jim Pollock, David Morrison, and Clark Chapman. All three collaborated with Sagan, contributing to papers, and continuing to be productive colleagues in later years—a tribute to Sagan's leadership ability in developing graduate students whom he advised. His students attributed this success to Carl's unique way of asking questions, always from outside the typical perspective.

Meanwhile, Fred Whipple, an expert on comets and meteors, began gently steering Sagan toward areas more likely than exobiology to produce solid, observable results. But Sagan had many interests that fell in the controversial realm besides life on other worlds. He also stated he believed interstellar travel was part of human destiny. In a speech to members of the American Rocket Society in Los Angeles in 1961, he had already surmised "other civilizations, aeons more advanced than ours, must today be plying the spaces between the stars" (Poundstone, 1999, p. 75). He also published a paper on "Direct Contact among Galactic Civilizations by Relativistic Interstellar Spaceflight."

Making use of the time warp afforded by the physics of relativity and forthcoming technology, he believed people would one day travel to far-off galaxies, returning to an Earth where tens of thousands of years have passed in their absence. (What an adventure! was Carl's take on this problem.)

Meanwhile, some of his colleagues acted more than a little snobbish about what they saw as a lack of seriousness in this supremely confident young man, a quality Frank Drake had alternately termed "brash."

But exobiology (especially) and interstellar space travel (less so) ranked among Carl Sagan's most burning interests. He explored these questions seriously and with an ever-increasing consciousness of the difference between wishes and facts that are bolstered by evidence. But his fertile imagination and eager curiosity were also among his greatest traits, very likely the source of his ability to ask productive research questions.

During this period, Sagan discovered Iosif Samuilovich Shklovskii, an astrophysicist from the Ukraine still living in the Soviet Union. Shklovskii was a kind of freewheeling idea generator, something like Carl, and in 1962 Shklovskii had published a theory postulating that a nearby explosion of a supernova about eighty million years ago emitted a deadly rain of cosmic rays that annihilated the dinosaurs.

Sensing that Shklovskii might be receptive to his freewheeling discussion of interstellar travel, Carl sent a cover letter and a preprint of his article (which included a discussion of the Drake Equation). The Cold War created many obstacles to free communication among scientists in the USSR and the United States, so Shklovskii may not have seen the Drake Equation before, and he was very interested.

From this overture a friendship and alliance opened up between the two, which resulted in an American edition of Shklovskii's book *Universe, Life, Intelligence*, originally published in Russian in 1962. Through his friend and mentor Joshua Lederberg, in 1961 Carl was appointed to the editorial board of Holden-Day, Inc., a newly formed publishing company specializing in science and located in San Francisco. As he and Shklovskii began to hash out the arrangements for the American edition,

Sagan used his contact with Holden-Day. A translation into English was completed and Shklovskii invited Sagan to make additions freely as he saw fit. The finished collaboration finally came out in 1966: *Intelligent Life in the Universe* by I.S. Shklovskii and Carl Sagan and published by Holden-Day. Sagan had taken Shklovskii's invitation to heart—his additions doubled the size of the book.

Sagan and Shklovskii completed the entire collaboration by mail, slowly and tediously waiting for letters to arrive. (E-mail could have made their exchanges so much easier!) For some time, they despaired of ever having the chance to meet—the Soviet government was reluctant to let Shklovskii out of its control. Eventually, they did meet, and one of the first times was in 1971 at the International SETI conference at the Byurakan Observatory in Soviet Armenia; it was the first time, in fact, most of the Soviet scientists in this field had a chance to meet their international colleagues.

MARS JARS AND MORE

During the 1960s, on the West Coast and at Harvard, Sagan also pursued biological research in several topics. He explored the possibility of microenvironments, a type of ecosystem both he and Lederberg thought might exist on Mars. Granted the conditions on Mars were inhospitable to life as we know it, but just possibly there could be tiny pockets or oases where life could develop and thrive. Today, NASA is using a search approach adapted to this view that life might exist in tiny underground or sheltered niches. Ice-covered lakes, which sometimes harbor life in the environmental extremes of Antarctica, are considered a promising location on Mars, as are hot springs. Looking for life or fossil life on Mars will probably involve digging.

Sagan also used a technique called "Mars jars" to simulate Martian ecosystems by placing organisms from Earth in "jars" that were subjected to Martian-like conditions. At a Denver conference called "Symposium on the Exploration of Mars," he reported that some of the microorganisms subjected to this simulation survived, even reproduced, showing that life may be possible on Mars, but not proving, of course, that life does exist there.

As development began in the 1960s for the Viking landers of the mid-1970s, scientists including Lederberg and Wolf Vishniac began to develop tests for the presence of life that a robot could perform. NASA hired Sagan to evaluate and select the tests to be used on the spacecraft. Much apprehension and curiosity accompanied this question. What might be seen hopping by a camera lens? Just in case, Sagan insisted upon including a camera, which turned out to be one of the best possible public relations moves for NASA. The only experience better than a picture of a faraway place is the experience of actually going there oneself—even if the picture shows nothing at all hopping by.

CAREER CRISIS

As Sagan's fifth anniversary at Harvard approached, the question of tenure arose. This weird aspect of academic life assures the holder of tenure a secure future (and ostensibly protects freedom of speech). However, for those who do not receive tenure, the issue becomes embattled. If one does not receive tenure, usually a contract may be offered on a yearly basis as before, but no assurance exists that the contract will be renewed from year to year. If one is on a tenure track and does not receive an offer of tenure, the impact on one's reputation is severe.

Sagan's tenure should have been fine. Whipple assured him there was no problem. Then suddenly like a bolt of lightning the unexpected and unwelcome news came: Tenure had been denied.

Why? Considerable conjecture has been made on this subject. Perhaps colleagues were jealous of Sagan's command of language and showmanship. Perhaps they were annoyed by his willingness to entertain ideas they might have considered outlandish. Perhaps they did not think his research topics were sufficiently substantive.

Also, recommendations from former professors and colleagues were requested by the deciding committee. Specifically, a recommendation was requested from Harold Urey, and Whipple found that Urey wrote a very unfavorable response. No one is quite sure exactly why.

Whipple wanted Sagan to stay and try again the following year, and Sagan did receive tenure from the Smithsonian Astrophysical Observatory, but he decided to move on. In 1968 he became Director of the Laboratory for Planetary Studies at Cornell University at Ithaca, New York. It proved to be a remarkably good move.

To young Carl and millions of others, the 1939 World's Fair, with its futuristic symbols of the trylon and perisphere, conveyed a spirit of optimism and a celebration of science and technology. Gottscho-Scheisner Collection. Courtesy of the Library of Congress.

Ekhart Hall, the building where Carl and Lynn met at the University of Chicago. Courtesy of The University of Chicago.

Above: Chandra, Kuiper, and Struve in front of the Yerkes Observatory. Yerkes Observatory Photograph.

Left: Tatel (85–1) Telescope used by Frank Drake in Project Ozma. By permission of the National Radio Astronomy Observatory/AUI.

Sagan with a model of the Viking lander. By permission of NASA/JPL/Caltech.

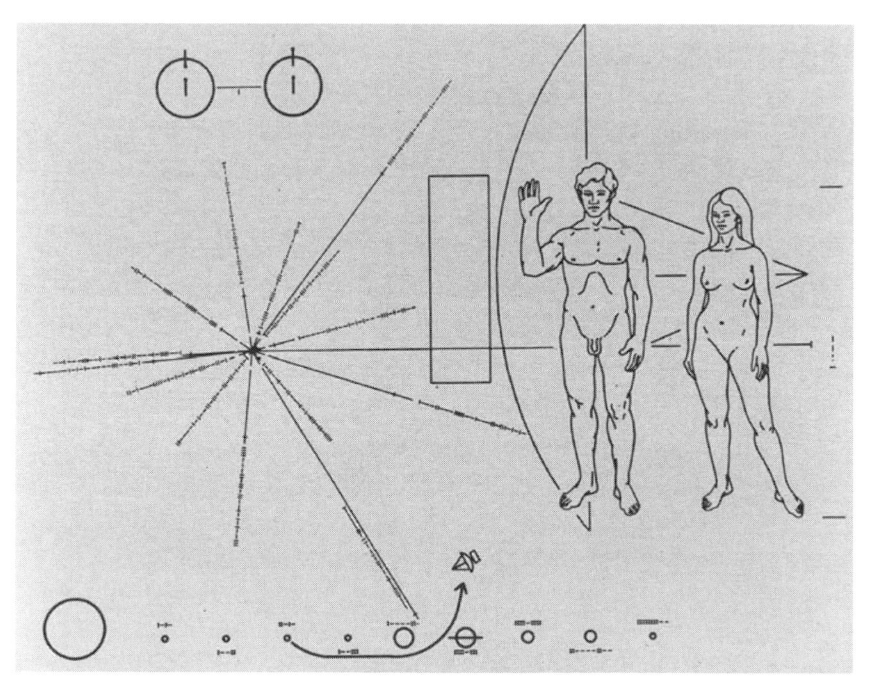

The Pioneer plaque. By permission of NASA/JPL/Caltech.

Sagan is among the founders of the Planetary Society. By permission of NASA/JPL/Caltech.

Sagan testifying for the space program. By permission of Bettmann/Corbis.

Radio telescopes at sunset. By permission of Photo Researchers, Inc.

Carl Sagan at the International Astronomical Union Meeting in Sydney, Australia, 1973. By permission of AIP Emilio Segrè Visual Archives. John Irwin Collection.

Chapter 6

MAKING COSMIC CONNECTIONS

When Carl Sagan and Linda Salzman met, he was by then dashing and single, witty and darkly handsome; she, an art student in the Boston area, was fun-loving, bouncy, voluptuous, and upbeat, with curly hair and a sense of style. More artistic than intellectual, she brought a new dimension of creativity to Carl's intense, cerebral life. But as he prepared to move from Cambridge to Ithaca, New York, Linda delivered an ultimatum. If Carl wanted her to come with him, it was time to get married. And so they did, on April 6, 1968.

ITHACA

Despite the considerable ego blow of being denied tenure by Harvard, Sagan may have done well to escape the stolid New England atmosphere that prevailed at Harvard. Ithaca was looser and more tolerant, with a hippie presence in the surrounding rural countryside and a more liberal outlook. Harvard was a magnet to decorated intellectuals, the stars of academia, and in Cambridge Sagan stood out mainly in the ways he was eccentric by Harvard standards. In Ithaca, he stood out as a star in his own right.

As a result of a connection Sagan made while consulting for the RAND Corporation in California, he had been selected in 1962 to sit on the editorial staff of the peer-reviewed journal *Icarus* (published by of the Division of Planetary Sciences of the American Astronomical Society). Now, in July 1968, an editorial slot opened up, and he became the natural choice for the job. Kuiper backed him for the position, and so did Joseph

Chamberlain of Yerkes, an astronomer who had once criticized Sagan's position on the Venus greenhouse effect, which had by now been vindicated by *Mariner 2*. So the editorial home of *Icarus* moved to Cornell with Sagan, providing one more feather in the Cornell cap. By January 1969, he was editor-in-chief, a position he would hold for more than a decade. It was a good time for it: A lot was happening in planetary science.

Meanwhile, Carl's achalasia had become an increasingly aggravating problem, and so in June 1969 he decided to undergo surgery on his esophagus to make swallowing easier. The surgery seemed to go well, and when his friend science-fiction writer Isaac Asimov visited him on June 18, they joked and planned to go to dinner after Carl's release. However, complications developed. Carl's lung collapsed and the lung cavity filled with blood. His friend Lester Grinspoon, a Harvard Medical School psychiatrist, stayed by his side throughout that night and the next, lifting Carl by his shoulders nearly twenty times an hour. The procedure allowed Sagan to breathe more deeply, as the hospital staff replaced nine units of blood and placed drains in his chest. Asimov also visited often and reassured Linda. After all that, the surgery did not correct his achalasia. The disorder continued to plague him most of his life.

As Carl recovered from his surgery and his brush with death, one of the landmark events in the history of humanity took place as *Apollo 11* astronauts Neil Armstrong and Buzz Aldrin landed and took their first steps on the Moon. The date was July 20, 1969. Blurry eyed from pain medication, Sagan watched on the TV set in his hospital room as the two moonwalkers left their footprints in the soft, powdery dust. It was a date that changed forever the parochial perspectives of humankind, allowing humankind to view Earth rising above the Moon's horizon. Years later, thanks to Sagan, humans would take a look at Earth from an even greater distance—from the orbit of Neptune, an image captured by *Voyager* looking back at our "pale blue dot" of a planet, revealed as tiny and fragile (Sagan, 1994).

A MESSAGE TO THE STARS

In December 1969 an exciting prospect came to light. Eric Burgess, who was a writer for the *Christian Science Monitor,* and Richard Hoagland, a planetarium lecturer, approached Sagan with an idea they thought he might like, given his reputation for having an interest in exobiology. In a few years, a spacecraft named *Pioneer 10* would begin a virtually endless journey into the cosmos. After leaving Earth it would swing by Jupiter to give us our first close look at the giant planet, take some measurements of

Jupiter's powerful magnetic field, and then, using Jupiter's gravitational pull like a slingshot, speed off toward the edge of the solar system, eventually leaving Sol (our Sun) and its nine planets behind to journey forever through interstellar space. What Burgess and Hoagland proposed was using this opportunity to send a physical message, a greeting card from Earth to whoever might be out there, beyond the edge of our solar system.

At a conference both Sagan and Frank Drake were attending, the two talked over the idea. (Even though the two scientists were now both at Cornell, with offices just down the hall from each other, somehow catching a moment of discussion back at Ithaca was harder than sitting down to talk at a conference.) Carl was galvanized. He checked with NASA to see if the agency would go along with the idea of affixing some kind of message to the spacecraft. The answer was "Yes!" *Pioneer 10* was about to become civilization's technological emissary in space. As for Drake, this was the kind of opportunity he had hoped for for years. The two scientists began brainstorming what form the message should take. They would have an area measuring about six inches by nine inches on a metal plate. In fact, there was room for line drawings. They immediately decided line drawings were a good idea. How about a galactic map? Great idea—but how should they ensure the map would still be usable even when, over aeons, the shapes of constellations change? Then Drake hit upon the idea of using a pulsar (an object in space that emits pulsating radio waves) to show the historical time and place of the launch. They could give the location of our solar system in terms of a couple of easily located pulsars. Since each pulsar has its own characteristic pulsing frequency, the pulsars should be easy for someone from another corner of the universe to identify, if the unknown readers were smart enough to capture the spacecraft in the first place. So the message would show a pulsar map of our Galaxy, the Milky Way. The pulsar map was Drake's job, including each pulsar's rotation period in binary numbers, the universal language of the universe (at least the best language anyone on Earth knew for communication with intelligent life from other worlds). Other information was included concerning the hydrogen atom, the simplest element in the universe.

Since Linda Salzman was an artist, the two planners co-opted her to do the line drawings. The finished drawing was etched on a gold-anodized aluminum plaque. Linda had drawn a nude man, his hand raised in friendly greeting, and a nude woman next to him. Trying not to offend any group of people, she intentionally blended their features so as not to single out any national or regional group as favored. She made both figures average in height. The job was finished, the plaque was securely attached in place, and on March 2, 1972, *Pioneer 10* was launched and began its long, long journey. Newscasters broadcast the plaque as the rocket lifted off the launchpad.

The outcry, which no one had foreseen, was tremendous. People complained about the nudity—both on TV and in space. Feminists complained the woman looked subservient. (Why wasn't she the one who offered the greeting?) Linda had meant for both figures to have brown hair, but since it was a line drawing with no shading, both figures looked like blonds. Salzman was a staunch feminist liberal (far more liberal than Sagan) and adamantly pro-Israeli, but otherwise about as unbiased a choice for the job as anyone could hope to find. Yet, NASA was accused of spreading smut and a prejudiced view of humanity at taxpayers' expense. Some critics conceded they could take solace in the fact probably no alien beings would ever see the pictures, because *Pioneer 10* was not heading toward any particular star.

There were some bright spots, though. In September 1970, Linda gave birth to a son, Nicholas Julian Zapata Sagan (soon known as Nick). A precociously verbal child, he immediately won his grandmother Rachel's heart. Also in 1970, Carl received a promotion to full professor of astronomy and space science at Cornell. And he began work that same year on JPL's Viking mission to Mars, of which Carl soon became the "guiding light."

SOJOURN TO ARMENIA

Sagan, Frank Drake, and the other members of the Society of the Dolphin (the name adopted by those who attended the first SETI conference) had long felt an urge to overcome the politically imposed separation and isolation from their colleagues in the Soviet Union, who also were actively launching searches for extraterrestrial intelligence. If only they could meet, compare notes, share methodology, brainstorm, and check their assumptions with one another.

In the Soviet Union, the king of SETI was a man named Nikolai Kardashev (1932–). A student of Shklovskii, Kardashev had been drawn to astronomy since early childhood, when his mother began taking him to the Moscow Planetarium. Fascinated by the stars and planets he saw featured there, he received his Ph.D. from the Sternberg Astronomical Institute at Moscow University in 1962 and quickly became a leading force among Soviet astronomers. That same year, Kardashev organized the first Soviet search for extraterrestrial intelligence, which involved the examination of an object catalogued by Caltech as "CTA-102." CTA-102 had first appeared to be evidence of another civilization—a supercivilization.

The finding created a worldwide stir, but CTA-102 turned out to be just a quasar, as shown by observations made from Palomar Observatory. Undaunted, Kardashev helped organize a Soviet conference on extraterrestrial intelligence at Byurakan Astrophysical Observatory in Soviet Armenia. The conference took place in 1964, three years after the Green Bank conference.

Kardashev theorized that superintelligent civilizations in the universe would be easily visible by their works, in other words by the ways in which they would have changed their surrounding planets, stars, and galaxies. Kardashev also proposed such superintelligent extraterrestrial civilizations would be able to harness all the power emitted by the sun central to their planetary system—or even harness the power of their entire galaxy. At the 1964 Soviet SETI conference, Kardashev proposed such civilizations might be billions of years in advance of Earth's current civilization.

A personable individual just two years older than Sagan, Kardashev appeared perpetually much younger than his years. He had energy and bold ideas, and his reports at the 1964 conference gained him international attention. Extraterrestrial civilizations might be divided into three types by their level of technological development, he conjectured. Type I would be those at Earth's level, that is, able to capture and use the resources of their own planet. He postulated two, more powerful levels. Type II civilizations would be able to generate roughly as much power as the Sun and would average some 3.2 more millennia in age. Type III would be able to generate levels of power the equivalent to an entire galaxy containing hundreds of billions of stars.

In a way, Kardashev was Sagan's Soviet counterpart, and the first international meeting on intergalactic communication was their brainchild, a collaboration between the two of them. Planned for 1971, and also slated to take place at the Byurakan Astrophysical Observatory, the conference was sponsored by the national academies of the two nations, organized by a team of scientists from each. From the United States, Sagan, Drake, and Morrison formed the nucleus, and from the Soviet Union they were Kardashev, Shklovskii, and two others, Viktor Ambartsumian, founder of the Byurakan Observatory, and Vsevolod Troitskii, who had begun an ambitious all-sky SETI survey in 1970.

So in September 1971, the Armenian observatory welcomed 28 Soviet participants, 15 American counterparts, and 4 from other countries. With the Soviet emphasis resting more on communication and broadcasting powerful signals to potential supercivilizations, and the American emphasis on listening, the mix made for interesting discussions. The topics of discussion ranged from radio search methods to what happens when you make contact, and a myriad of topics in between. Using Drake's by now famous equation, the conference tentatively concluded as many as a million civilizations might exist in the Galaxy (the Milky Way) that would be capable of communicating across the great distances between us—that is, that would be our technical equals or more sophisticated. But, of course, everything depended on the answers to each question. The "softest" answer of all—that is the one fed by the least objective, least factual numbers—was the value of L. No one knows how long a technologically sophisticated civilization can last. The twentieth century brought with its innovations the potential for communicating with extraterrestrial civilizations, but it also brought the means for self-annihilation. How many civilizations might survive that grim temptation (not to mention natural disasters such as an asteroid collision)? This is a question that had haunted Sagan for years and would become the impetus for activism later in his life.

THE COSMIC CONNECTION

The year 1973 saw the publication of Sagan's first book, *The Cosmic Connection: An Extraterrestrial Perspective*. A celebration of science, astronomy, and the universe, this book also began to show a polemic side of Sagan that would come out more strongly in his later years. This was Sagan developing teeth. As he revealed perspectives on the universe, its past, present, and future, he also held up a mirror to the way we think about ourselves and our society and our place in the universe. In the process, he showed clearly that his political consciousness was beginning to awaken as he took on the military, following up on an essay published in 1971 that strongly criticized bombing raids on Vietnam.

The book came into being as the vision of an entrepreneur named Jerome Agel, who made a fortune backing the work of Canadian writer Marshall McLuhan, a sort of media and pop-culture guru of the 1960s. Agel had produced McLuhan's runaway bestseller, *The Medium Is the Massage: An Inventory of Effects* (1967), after which Agel proceeded to produce a succession of books on popular topics. A veritable fountain of ideas for publications, Agel would seek the perfect author, usually someone with both the relevant expertise and a degree of fame. Like Sagan, as a child Agel had dreamed of the planets and particularly of Mars. Now he conceived of a book that would celebrate the cosmos, even the possibility of life on other worlds. After reading an article on Sagan in the *New York*

Post Magazine, he contacted Sagan and set up a meeting. The two men immediately saw eye to eye.

The topic, though, had become unpopular. The United States continued to send missions to the Moon through December 1972, but a bloody war in Vietnam, riots and unrest at home and, perhaps, a short attention span had turned public interest away from NASA and things cosmic. The two talented persuaders offered their book proposal to 17 publishers without a nibble. Finally, they landed a contract with Doubleday, earning Sagan an advance of $18,000 (a kingly sum, which would have a purchasing power of about $80,000 today).

For the book, Sagan wrote a series or short essays in an easy going, personable style on a wide range of topics. Sagan was rarely afraid of dreaming about possibilities, and his lyrical prose evoked an excitement about the solar system and the universe that droning, staid scientific data could not convey to the average reader. Sagan set real scientific information out in a way that provided context. He daringly imagined future possibilities with titles such as "Terraforming the Planets," in which he poses the possibility that humans could reshape the environment of Mars or Venus—so clearly inhospitable to life—so that it would become habitable. In "Some of My Best Friends Are Dolphins," Sagan recalls John Lilly's enchanting tales about his dolphins and speculation about their intelligence. In "Hello, Central Casting? Send Me Twenty Extraterrestrials," he entertains some possibilities of what beings from other worlds might be like. All of this was daring, and some of it attracted negative critiques by scientists. But Sagan was drawing his audience in. His picture—a young, handsome man, with a dark shock of hair, an amiable smile and a casual turtleneck pullover—contrasted favorably in the early 1970s with the "suits" that a large contingent of teens and college students mistrusted. He looked like someone you could talk to, someone who might be open to unconventional ideas. His writing in these speculative essays confirmed the assumptions. Sagan was open-minded, willing to look at possibilities. The scientist in him was critical and looked for proof. He argued objectively, and he let go of disproved ideas. But he did not reject unproven ideas out of hand.

But Sagan also wove some debunking into this book. While holding out for the reader the wonders of the universe, he argued against scientifically insupportable ideas such as those set forth by Immanuel Velikovsky in his highly popular book, *Worlds in Collision*, or by Erik von Danniken in his *Chariots of the Gods*. Both still have fans today who defend them ardently.

Velikovsky proposed that many ancient myths and legends are based on actual visitations to Earth made by extraterrestrial spacecraft. He based his account of ancient celestial wars on age-old manuscripts and drawings, old astronomical inscriptions, and other evidence. For instance, he cites ancient cave drawings of beings having elongated heads that look like space helmets. However, as Sagan points out, "In fact the expectation that extraterrestrial astronauts would look precisely like American or Soviet astronauts, down to their space suits and eyeballs, is probably less credible than the idea of a visitation itself" (1973, p. 206). Again, Occam's razor is brought in to trim off the unnecessary excess. The artist may have been inept. Perhaps the drawings portray ceremonial masks. And so on.

Van Danniken also claimed ancient astronauts visited Earth. Markings found high in the Andes were ancient landing strips, he asserted, where they landed their rockets. These visitors also introduced nuclear weapons and other artifacts. Sagan dismissed these ideas as "implausible" and even "unimaginative" because they mimic the recent history of human technological development so closely. "Most popular accounts of alleged contact with extraterrestrials," adds Sagan, "are strikingly chauvinistic" (1973, p. 208).

Sagan was on his way to becoming science's most persuasive spokesperson against pseudoscience, precisely because he did not look or act the part. His mixture of speculation and debunking, combined with imaginative illustrations peppered throughout the book, made for a winner. Sales on the hardcover edition were so stunning that Dell purchased paperback rights from Doubleday for $350,000.

Agel revved up the publicity blitz. Sagan spoke at meetings, wrote articles, gave interviews, and appeared on television, including a 30-minute appearance on CBS-TV's *Camera 3*. One evening in the early 1970s, TV's *Tonight Show* host, Johnny Carson, caught an appearance by Sagan on rival late-night talk-show host Dick Cavett's program. Carson contacted Sagan and invited him to appear on the *Tonight Show*. His first appearance was November 30, 1973. The evening was a roaring success. Unlike Robert Jastrow, who had been Carson's "house astronomer," Sagan did not wear glasses or look academic or seem distant. Sagan had energy, humor, and wit, an easy conversational manner, and always, the utmost self-confidence. His descriptions and explanations were vivid and colloquial. He had all the best qualities of a good travel writer. When he spoke about Jupiter or Mars or Venus, listeners felt as if they had actually gone there with Sagan as tour guide. The audience loved it. That night began a 20-year relationship between Carson's show and the new on-call astronomer, Carl Sagan.

Time magazine named 200 "rising leaders" of young adults in its July 15, 1974, issue. Carl Sagan was on the list. *TV Guide* ran a piece, read by some ten million readers, timed to appear the same week as Sagan's appearance on the Public Broadcasting System (PBS) series *Nova* in a program called "The Search for Life." The *TV Guide* article, entitled "Seeking the Cosmic Jackpot," included enthusiastic comments from Sagan, declaring that finding life on Mars would be a "cosmic jackpot." Such a coup, he promised, would net us a gain in "biological perspectives" and "practical benefits for mankind," including a rekindling of our human sense of wonder and adventure. NASA could not have said it better and today's exobiologists could not agree more.

During this period of active publicity, a talented young science journalist named Timothy Ferris got the bright idea that Sagan might be a good interview subject for the quintessential counterculture publication *Rolling Stone*. The article idea was a stroke of genius all the way around. Sagan proved to be a great interview, an enthusiastically liberal, engaging figure who exhibited socially relevant values and counterculture mores in the unexpected guise of a scientist, of all things. At a time when folk singers wrote acrid antitechnology lyrics performed to sold-out crowds, and activists by the thousands marched in the streets against warfare and the establishment, the match seemed highly improbable. But Ferris, who at the time was New York Bureau Chief for *Rolling Stone*, has a knack for wrapping science in metaphors that people understand. Like Sagan, he is a polymath, grounded in the richness of music, literature, dance, theater, history, and philosophy. He ignores and slips past what one biographer calls the "firewalls of thought" (Weed, 2000) and counts as his friends individuals as varied as gonzo journalist Hunter S. Thompson, and literary novelists Joyce Carol Oates, Jim Harrison, and Annie Dillard.

The article appeared on June 7, 1973, greatly enhancing Sagan's position as, one might say, a scientist of the people. Sagan had a warm, relaxed demeanor that helped to offset the conventional image left by scientists of the ivory-tower persuasion who scorned those who did not understand their work, their methods, or their priorities. Carl Sagan seemed more like a regular guy.

Sagan and Ferris became friends. With frequent meetings with publishers, TV appearances, and other obligations in New York, Sagan found he often needed a temporary home base in Manhattan, and Ferris generously offered his apartment. Ferris was working on his forthcoming book on astronomy, *The Red Limit: The Search for the Edge of the Universe* (1977, 1st ed.)—for which Sagan would write the introduction, and magazine pieces—but Ferris was also still working for *Rolling Stone* and had lots of interesting acquaintances and connections, among them writer/director Nora Ephron. Sagan sometimes tagged along to parties and gatherings, where his love of conversation—both holding forth and listening attentively—made him a welcome guest.

One such gathering in the fall of 1974 became particularly memorable. Ephron held a small dinner party, and the list of invitees included Carl and Linda Sagan; Timothy Ferris and his girlfriend Ann Druyan, who was working on a novel, *A Famous Broken Heart;* writer Lynda Obst and film producer David Obst, who was then Lynda's husband; Frank Rich, a critic; and historian Taylor Branch. When Druyan arrived, Sagan was clowning around, laughing uproariously stretched out on the living-room floor. At 25, Druyan was some 15 years younger than Sagan, but the difference in years seemed nonexistent. She was intelligent, intellectually curious, upbeat, and had a wonderful sense of humor and eclectic interests. Sagan and Druyan became friends practically on the spot, especially when Druyan began talking about her love for baseball. She could quote stats for any league, major, minor, or Negro. She had visited the Baseball Hall of Fame in Cooperstown, New York. She hailed from Queens, New York, loved good books, read Mark Twain by flashlight when her parents thought she was sleeping, and majored in literature at New York University.

SETI AT ARECIBO

As Sagan's 40th birthday approached, he finally succeeded in lining up an opportunity to do a hands-on SETI search with Frank Drake on the 1,000-foot dish at Arecibo Observatory, a radio telescope installed in a natural bowl-shaped hollow about nine miles south of Arecibo, Puerto Rico, on the northern coast. At the 1971 SETI conference at Byurakan, Drake suggested using of the Arecibo dish to conduct a SETI search of nearby galaxies. Ever since Byurakan, Sagan had been urging Drake to book the time and set up the search at the facility completed by Cornell in 1963 and operated under contract with the National Science Foundation. "Let's do it!" he would say with a twinkle in his eye whenever he would run into Drake on campus in Ithaca. "Let's do it!"

Now, in November 1974, Arecibo had just completed a major upgrade, and the time was perfect. A book entitled *The Galactic Club* by Stanford radio astronomer Ron Bracewell had just come out, postulating the existence of communities of advanced civilizations already in communication with each other. Perhaps they would be detectable by the narrow-beam beacons they might use to communicate with each other.

Another publication, known as the Project Cyclops report (written primarily by Barney Oliver, one of the original members of the Order of the Dolphins), envisioned a day when Earth might become a member of such a galactic-wide supercommunity. (Project Cyclops was a scheme for linking fifteen hundred radio telescopes to act as a single eye—or rather, ear—with enormous sensitivity to remote signals. It was never funded, however.)

Nearly everyone in the SETI community expected that when contact was made, the extraterrestrials would be more advanced than humans. No civilization with less technological expertise than we had would be able either to send or receive signals. That process required a knowledge of radio waves and radio equipment. The possibilities for more expertise than we had, however, were almost endless.

Sagan and Drake planned to search four galaxies located just beyond the Milky Way. The equipment installed at Arecibo allowed for scanning enormous numbers—hundreds of billions—of stars at one time, a vastly more efficient procedure than Drake had been able to use with Project Ozma. Also, using the newly installed multichannel receivers, they could pick up a thousand channels at once. They figured that if even one super-civilization existed in one of those galaxies, Arecibo would pick up its signal. Of course, it would be so far away that its signal would have left home millions, perhaps billions, of years ago. But if a powerful, bright signal was out there aimed in the direction of the giant dish nestled in the quiet Puerto Rican valley, they would find it. Maybe even if they were not trying to communicate but were emitting radiation in some recognizable repeated pattern, even if they were broadcasting unintentionally, they would find them. For example, Arecibo sends out a signal that is ten million times brighter than the Sun.

No accommodations were available for the two astronomers at the observatory, so they stayed in a small seaside hotel and drove up to the observatory early in the morning, when their target galaxies rose above the horizon. It was a long drive on a rough, narrow road, so they got up before dawn and munched garlic bread on the way. Once there, they were caught up in the process of watching the monitors for intentional, intelligent patterns and of steering the big antenna. Meanwhile the Sun rose, spreading rosy light across the dawn skies. Watching the green points of light on the oscilloscope, they monitored, while the telescope scanned more area in a hundredth of a second than Drake had been able to cover in two months at Green Bank.

They began with M-33, also known as the Great Nebula in Triangulum. (M-33 is the number assigned by eighteenth-century astronomer

Charles Messier in his catalogue of nebulous objects.) For Drake, watching the scope and monitoring the search came as second nature. Sagan was also present with Drake in the control room, but for him it was a restless, unproductive-feeling experience. As Drake later explained, Sagan was used to more activity and more tangible results. "When he sat in the control room at JPL, phenomenal photographs from other worlds would stream in, another one every couple of minutes" (Drake & Sobel, 1992, p. 151). Not so the control room at Arecibo, which was quiet and lonely and uneventful.

The first half hour went by and, 10 billion stars later, they had seen nothing promising. After a full hour, still nothing. "I could sense Carl's disappointment. After a few days he was even a little bored by the sight of the green dots appearing uneventfully on the screen. And who could blame him?" Drake also admitted to feeling some disappointment, but, he explained, he was "just more used to a slower unraveling of the universe's mysteries." Even a hunt for pulsars or mapping the radio sky would give more satisfaction, though. This was a different kind of astronomical search and would appeal to a very particular kind of patient, persevering personality. The reward would be that incredible moment when an undeniably intelligent pattern does appear on that screen. That moment never came though, for Frank and Carl in the hundred hours they spent watching the monitor, checking the data, and waiting.

VIKING AND THE SEARCH FOR LIFE ON MARS

Between 1965 and 1971, NASA sent four successful missions to Mars—*Mariner 4*, *Mariner 6*, and *Mariner 7* each flew by after brief encounters. *Mariner 4* was the first successful mission to Mars and therefore provided the first information gathered close-up. The spacecraft arrived at the red planet July 14, 1965. Among other measurements, it reported that no radiation belts and no magnetic field were present, which told planetologists Mars has no metallic core. But to Sagan and other scientists hoping to find signs of life on Mars, *Mariner 4* was a disappointment. Its measurements showed the pressure of the Martian carbon dioxide atmosphere was only 1 percent of the atmospheric pressure on Earth, making the existence of Earth-like life more unlikely. Also, unlike the first U.S. probe to Venus, *Mariner 4* carried cameras. But the images showed nothing very exciting, twenty or so blurry images of a dismal expanse of rust-colored, desert-like surface covered by craters. No sign of the famous canals Percival Lowell had reported seeing through his telescope earlier in the century—long since discredited but not completely dismissed by some people. No sign of any civilization at all. In fact, nothing that fostered hope of finding any life at all. Mars appeared to be as lifeless and desolate as our own Moon.

Mariner 6 and *Mariner* 7 composed NASA's first tandem mission to Mars, flying by during midsummer 1969. Between them, they captured more than 126 images and used radio, ultraviolet, and infrared instruments to study the surface and atmosphere.

But *Mariner* 9 was the most intriguing of the early spaceflights to Mars. Launched May 30, 1971, it arrived at the red planet five and a half months later, on November 13, when it entered orbit. *Mariner* 9 thereby became the first artificial satellite of a planet other than Earth. This innovation allowed for much lengthier and more complete observations and reports, including more than seven thousand photos. At first, however, the results were enormously disappointing, a sort of déjà vu, for Sagan, of his summer in Texas with Kuiper: *Mariner* 9 had arrived in the midst of another giant dust storm, and the first pictures showed hardly any surface features.

However, once the storm cleared up, the *Mariner* 9 photos were exciting. The spacecraft succeeded in mapping 85 percent of the planet's surface and sent back photos of the two small, potato-shaped moons of Mars, Phobos and Deimos. This was a real bird's-eye visit to Mars. For the first time, scientists saw the majestic Martian mountain dubbed Mons Olympus, the most massive known mountain in the solar system, with a height of 16.8 miles and a base so large it would cover the entire state of Missouri. *Mariner* 9 introduced them to the vast canyon Valles Marineris, which cuts across the face of Mars in a great gash nearly as long as the distance from New York to Los Angeles and three times as deep as the Grand Canyon. The stalwart orbiter also captured images of large regions showing evidence that extensive floodwaters and rivers may have flowed there in the past, providing perplexities of great interest to planetary geologists and a hopeful sign to exobiologists like Sagan.

By the mid-1970s, scientists and engineers were nearing another key launch date, twin spacecraft called *Viking 1* and *Viking 2*. Each spacecraft was composed of two parts: an orbiter that would map and take measurements from orbit and a lander designed to descend to the surface, photograph, and run tests, including four tests designed to test for the presence of life or organic substances in Martian soil.

A member of the Viking landing-site team, Sagan took a leave of absence from Cornell in 1976, and he, Linda, and Nick spent a good part of that year living in an apartment in Pasadena, California, where JPL and Caltech are located. Ann Druyan and Timothy Ferris were also there.

Always eager to pique the interest of listeners and readers, in his many speeches and articles Sagan speculated openly about what the Viking landers' cameras might reveal. Creatures living on Mars were not necessarily small, he asserted. Given the extreme cold, the low atmospheric pressure, and the harsh UV radiation from the Sun, he pointed out, larger creatures might benefit from the smaller ratio of surface area to total body volume and therefore be favored by natural selection. He talked about catching a view of creatures hopping by the cameras. And he lobbied for attaching a night-light to the lander, in case night creatures might visit the site under cover of darkness. Other members of the Viking landing-site team overruled him, though, and were sometimes chagrined at what they feared might be buildup of expectation. They worried that Sagan's careful cautionary phrase, "I don't think we can rule this out," was a subtlety the general public would not hear and disappointment could be just around the corner.

If Sagan was possibly more inclined to play up imaginative possibilities than were his colleagues, he was all business about the process of choosing a landing site. The initial decisions about where to land were based on compromises. The team was made up of scientists from different disciplines, and, of course, their objectives were different. Geologists hoped to see interesting land formations, and among that group there were those who were more interested in volcanoes, others in outcroppings, or revealing evidence of flowing water. Exobiologists wanted a region where life might most likely be evident. Some scientists wanted to know more about the polar region. But everyone conceded the landing site had to be safe for the lander, which would not be able to gather any information for anyone if, for example, its antenna was injured on landing. Based on *Mariner 9* photos, the team had settled on what looked like two open plains.

The two Viking spacecraft were launched in August 1975, with the Sagan family and many of the team in attendance at liftoff in Florida. A problem with the *Viking 1* battery caused a swap in the order, launching *Viking 2* before *Viking 1*, so there would be time to fix the problem. In the process, the two spacecraft also swapped names. Finally, both Viking spacecraft were safely on their way to the red planet.

As the lead spacecraft approached its destination, though, a new problem arose. The orbiter was maneuvered into position to take pictures and radar scans of the planned *Viking 1* landing site, and what the landing-site team saw made beads of perspiration appear on their brows. What had looked like a smooth, featureless plain in the *Mariner 9* images now turned out to be strewn with boulders—not a good place at all for a landing. Perhaps the dust storm that hampered *Mariner 9* was to blame. The search immediately began for a better site, using two means of evaluation: photo imagery and radar scans. The problem was that most possible sites appeared to be good when the team looked at the radar readings, but looked like trouble in the photos. And those that looked good in the photos showed information in the radar readings that probably would mean a risky landing. Some members of the team trusted radar over photos; others liked to see what they were doing and preferred the photos. Sagan pointed out that either approach held undue risks; instead, they should consider only sites that looked workable using *both* radar scans *and* photographs. Originally planned to land on July 4, 1976, the day of the U.S. bicentennial celebration, now the *Viking 1* landing was delayed. The team worked for days practically nonstop, determined not to lose the long-awaited Viking view and analysis from ground level. The walls of their conference room were covered with maps, some of which they spread out on tables, around which they clustered. Finally, the choice for *Viking 1* narrowed down to an area of the Chryse Planitia (Plains of Gold) near the original site. One geologist, Hal Masursky, thought the safest area would be dead center in the middle of the plain. Others didn't like the look of the center on the radar scans and preferred areas closer to the edge. After obtaining and comparing a crater count in the three most favored landing sites, all of which looked comparatively good in both radar scans and images (but none of which were without risk—that was impossible to find), the consensus finally went to a compromise area halfway between the center and the edge. The decision was made.

The *Viking 1* lander made its separation from its orbiter and sped toward the red planet's surface. Right on schedule, about nineteen thousand feet above the surface, the parachute popped out, unfurled, and slowed the fall. Retrorockets kicked in to further slow the descent at about four thousand feet. *Viking 1* jutted out its legs and settled to the surface. The first spacecraft had landed on Mars. The date was July 20, 1976.

As the first photos came in from Mars, Sagan was on-camera providing commentary. He was in his element. The team had known the risks to the spacecraft and had planned each photo with care. The first color photo to come in showed a vast, bleak, rusty-red plain stretching out to the horizon. Rippling dunes of red sand were strewn with rocks and boulders beneath an extended sky, not black or blue as expected, but a surprisingly light, salmon color (presumably tinged by sunlight reflecting off reddish motes of dust floating in the thin Martian atmosphere). Several of the surrounding boulders could have toppled *Viking* and rendered it useless. It was a lucky landing.

Lacking completely was any sign of life. As Sagan put it, right away, "There was not a hint of life, no bushes, no trees, no cactus, no giraffes, antelopes, or rabbits." His disappointment didn't show, but that night he pored over the photos. Also there helping were Ann Druyan and Timothy Ferris. Perhaps they had so far missed some telltale sign. Could they find a burrow hole, a lichen-like growth on a rock, an artifact left behind by some creature, footprints, or some other shred of evidence. Was that an oasis in the distance? Rocks took on familiar shapes as clouds do sometimes on a summer day. But in the end there was no oasis, no auto parts or shoes, nothing but an expanse of desolate dirt and rock.

Meanwhile, a decision needed to be made about the *Viking 2* landing site. Hal Masursky reminded the team they had agreed if caution paid off in the case of *Viking 1*, they would go for a more interesting though possibly more risky site for *Viking 2*. But, like a basketball player who has just made the first of two free throws, the team members had a case of nerves.

As Sagan flipped a coin, he asked whether coming up heads once was any guarantee for coming up heads twice. Further, some 30 percent of the sites they had considered the safest had turned out to have dangerous boulders. Sagan and several other of the team members felt these facts did not really recommend raising the bar on risk. But Lederberg thought the farther north they went, the more likely they would find what they were looking for.

The lower latitudes were hot and dry and high. Not a good bet, Masursky concurred. Masursky asked the team what they would be likely to learn from a spacecraft located in each of the three major sites. The first site, called A-1, was the spot where a complex of channels opened out onto Chryse Planitia, making this a potentially ideal spot to look for signs of the past presence of near-surface ice or vast amounts of water, as well as complex organic molecules.

The second site, B-1, was located in an area where evidence indicated there might be a high concentration of water vapor. It was also a good location from the point of view of easy communication with both orbiters. However, it was outside the region available for radar scans, so the team would have no radar support in refining their site plans.

Both geologist Harold Klein and Sagan's friend, Josh Lederberg, believed the northern latitudes presented better odds for finding water. However, Vance Oyama pointed out that freezing temperatures prevailed in these regions. Lederberg finally won Sagan over to the idea of going north. By the 21st of July, after taking a close look at the first *Viking 1* pictures of the surface, the team members cast their votes: 3 for a site in the

C region, and 40 for B, to the north. From there, they quickly narrowed it down, and a few weeks later, on September 3, the *Viking 2* lander touched down safely in that area, known as Utopia Planitia (Plains of Utopia).

LIFE—ON MARS?

The Viking landers each carried three biology experiments designed to find out whether life existed on Mars. They did discover some chemical activity in the soil, which remained unexplained. However, no clear evidence emerged indicating microorganisms lived in the soil at either site. Mission biologists came to the conclusion that the Martian surface self-sterilizes and is therefore enormously hostile to life. That is, environmental factors actually prevent the formation of living organisms. Those factors include the harsh UV radiation from the Sun that permeates the planet's surface, the soil's extreme dryness, and the oxidizing process that takes place in the soil. But to some exobiologists the results seemed inconclusive. Some scientists continue to hold out hope that some remnant, some fossil, some microenvironment will ultimately be found in which there is some sign of at least microbial life, past or present.

The fourth on-board experiment, the gas chromatograph and mass spectrometer instruments on both landers, found no sign of organic chemistry. The instruments did produce an excellent analysis, however, of the Martian atmosphere's composition. Also, the X-ray fluorescence spectrometers measured the elements found in the Martian soil.

As for the presence of life in the ancient past of the planet, the jury on that question is definitely still out.

The two Viking landers—the first to land successfully on the surface of Mars—were also the only missions to land there for the next 20 years. No mission followed up Viking successfully until 1997 (*Pathfinder*, the spacecraft, and *Sojourner Truth*, the rover), after Sagan's death. With this mission, Sagan's wish for a mobile lander came true. *Sojourner* rolled about from one rock to another around the landing site, examining and testing. Signs of life remained elusive, though. After that mission another hiatus occurred—due to unsuccessful missions by NASA, Japan, and the European Space Agency—until the successful twin rovers *Spirit* and *Opportunity* arrived in early 2004.

THE DRAGONS OF EDEN

Even as some faculty members back in Ithaca were mumbling about Sagan's frequent and sometimes extended absences, Cornell further honored him. The year *Viking* landed, Sagan was appointed to the David Duncan chair as Professor of Astronomy and Space Sciences. His colleagues in astronomy did not tend to complain, however. They were well aware of the prestige and stature his "extracurricular activities" brought to their institution.

Things were not going so well for him at home, however. Linda and Carl had started out their relationship as an attraction of opposites. She was fiery and intuitive; he was cool, reasoned, and logical. His application of reason to matters of passion and emotion infuriated her. Their son Nick would later observe their diametric opposition, seeing himself as a mixture of the two and wondering why they ever got together. Like the Cornell faculty and Lynn Margulis as well, Linda complained he was never home and everything else took precedence over his family. In the early years at Ithaca, Linda enjoyed cooking gourmet meals for family and guests, smoothing Carl's social interactions at parties and get-togethers, and acting as the artistic leavening Carl seemed to need. But, as his schedule became busier and busier and as he pursued the intellectual and public life that had begun to unfold as his internally driven destiny, they grew farther and farther apart. Like Margulis, Sagan's second wife also complained of frustration that Carl left her with all the household and child-raising responsibilities with no regard for her own career.

As eclectic as ever, Sagan continued to explore the wide range of questions and topics that attracted his interest. In his next major publication, *The Dragons of Eden: Speculations on the Evolution of Human Intelligence*, published in 1977, he explored the brain, human intelligence, and how they may have evolved—a subject clearly related to questions about the prospect of intelligent life on other worlds. Is intelligent life a natural evolutionary path for life everywhere to take? What exactly is intelligence? How does it relate to the development of the human brain? It was also an outgrowth of Sagan's effort to understand the wedge driven between him and Salzman.

Although the book would not save his marriage, it met with immediate and widespread popular appreciation. Sagan had succeeded in celebrating the human brain in a way that engaged readers who probably thought they had little use for what their childhood schoolmates would have called "being brainy." "Provocative... entertaining... impressive...," ran the review in the *Chicago Daily News*. "Even those of us who are not heavy-weight thinkers will wish to pillow our craniums a bit more carefully at night after reading Sagan. He makes you realize that the gray mass between your ears is quite a treasure." His good friend Isaac Asimov wrote,

"Never have I read anything on the subject of human intelligence as fascinating and as charming."

Sagan concluded his book by making a case against pseudoscience, which he approached tough-mindedly, as always, while at the same time showing an appreciation for its attractions. He decries the turn away from scientific knowledge, which was already taking hold and grew more pronounced in later decades of the twentieth century. He compares the trend to the period known as the Dark Ages in Europe, when curiosity and the thirst to know were seen as "diseases" in the terminology of St. Augustine of Hippo. He classes such "diseases" as "antiscience," a large array of doctrines against which he cautions, declaring them "intellectually careless," lacking in tough-mindedness, and as "vague, anecdotal and often demonstrably erroneous" (Sagan, 1977, p. 247). The list includes astrology, the Bermuda Triangle, ancient astronauts, UFOs, auras, Velikovsky's colliding worlds, and spiritualism, to name a few. At the end of his list he adds, "the doctrine of special creation, by God or gods, of mankind despite our deep relatedness, both in biochemistry and in brain physiology, with the other animals" (p. 248). As always, Sagan softened his critique with a voiced understanding of the human need to reconcile emotional needs with the intellectual demands of today's overwhelmingly complex environment. But he cautions these more or less purely right-brain responses to life evoke mystical and occult underpinnings that do not respond to evaluation by left-brain standards. They cannot be disproved. They do not lend themselves to rational discussion. Functioning exclusively on intuition does not embrace the fullness of what it means to be human, Sagan would argue. Instead, he concludes, "In contrast, the aperture to a bright future lies almost certainly through the full functioning of the neocortex—reason alloyed with intuition and with limbic and R-complex components, to be sure, but reason nonetheless: a courageous working through the world as it really is."

In 1978, this remarkable book won the Pulitzer Prize for general nonfiction.

RECRUITING FOR THE CAUSE

He was also getting more involved with students. Carl made it his business to recruit strong students to the cause of science, and Cornell sometimes passed likely applications for admission on to him. Such was the case with a young man from a tough neighborhood in the Bronx in New York City who later described his application as "dripping with interest in the universe." As he tells the story, a few weeks later, a letter from Sagan arrived, inviting the young high school student to visit him in Ithaca. "Was this, I asked myself, the same Carl Sagan that I had seen on Johnny Carson? Was this the same Carl Sagan that had written those books on the universe? Indeed it was. I visited Carl on a snowy afternoon in February (I later learned that many winter afternoons in Ithaca are snowy). He was warm, compassionate, and demonstrated what appeared to be a genuine interest in my life's path. At the end of the day, he drove me back to the Ithaca bus station and jotted down his home phone number—just in case the buses could not navigate through the snow and I needed a place to stay."

Neil deGrasse Tyson attended Harvard instead, but Carl had made an indelible impression. Carl wrote a jacket blurb for Tyson's second book, *Universe Down to Earth*, and later invited Tyson to attend a special workshop called together by Dan Goldin, head of NASA. The purpose was to discuss the future of NASA's relevancy to the heart and mind of the American public. Tyson was the youngest person there. With that meeting Tyson became part of the cadre of scientists who set out to present science to the public using any tools at hand—articles, books, television, public lectures, teaching. Today Neil deGrasse Tyson is the Frederick P. Rose Director of New York City's Hayden Planetarium and a visiting Research Scientist and Lecturer at the Department of Astrophysics at Princeton University. He would later remark, "I never told him this, but at every stage of my scientific career that followed, I modeled my encounters with students after my first encounter with Carl" (Tyson, 1997).

THE VOYAGER MESSAGE

Meanwhile, another pair of spacecraft was getting ready to head through the outer planets and out of the solar system. This time the decision was made to place a record onboard that had been specially designed (before DVDs or photo CDs existed) to carry images as well as sounds of Earth and music. Sagan and Druyan teamed up to pick the music, sounds, and photos. Both felt guilty about the attraction they felt for each other, but both were feeling closer and closer. Finally, on June 1, 1977, they admitted their love to each other. Carl told Linda he had fallen in love with someone else. Linda was furious. Carl's parents met Annie, as everyone called her, and they both loved her immediately. When Carl told his father he was getting a divorce because he had fallen in love, Sam Sagan said, "I hope it's Annie."

A dozen years of Sagan's life had passed since he had left Harvard for Cornell. During that time he survived a close brush with death, began a new family, found new friends, and built a career. He helped spearhead national and international efforts to discover life on other worlds, pursued experimental efforts to simulate the beginnings of life on Earth, and was a key player in the exploration of the solar system.

During this period, Sagan also discovered he was exceptionally good at communicating, and he wrote highly acclaimed books that, with his TV appearances, established his name as a household word. More importantly, he opened the laboratory windows of science to show the rest of the world what science is really about. He had changed the way many people viewed science, blowing the dust off the laboratory shelves, polishing the beakers and test tubes, taking advantage of new technology to take people closer to the wonders of the stars and planets. He began to shine a spotlight on the exciting universe as seen through the eyes of science. And on the methods of science as a reliable way to seek knowledge.

The next few years in his life would build another, especially eloquent chapter in this effort, in what would prove to be one of Carl Sagan's key lifetime legacies, a visit to the cosmos. It would also prove to be the beginning of the most productive, satisfying, and happy partnership of his life, on both personal and professional levels.

Chapter 7

COSMOS: A SMASH HIT FOR SCIENCE

By 1978, Sagan had reached a new level of fame. His file drawers began to fill with communications he and his secretary Shirley Arden discreetly referred to as "fractured ceramics," a gentle euphemism for "crackpots." Would-be inventors and theorists called to talk to him or sent him inch-thick manuscripts, and fans dropped into his office to visit or play ping-pong, as one young man proposed. Protecting Sagan's time became a job in itself.

But in addition to bringing nuisance-mail, fame can also open doors and create opportunities. Seeds planted just a few years back in chapters of *The Cosmic Connection* began to sprout. Sagan's lyrical prose on space exploration and its value evoked such memorable wonders and transported readers so successfully into the vast and extraordinary reaches of the universe that critics and readers alike began to see Sagan as a sort of prophet, in the sense that he seemed to hold the keys allowing nonscientists into these spectacular realms to experience the incomparable beauty that scientists reveal through their work.

Originally, the idea for *Cosmos* came from Gentry Lee, who served as JPL Director of Science Analysis and Mission Planning on the Viking missions. Both the public and the press had reacted so inconsistently to the Viking missions, with their interest first running hot and then cold, that the scientists were left feeling jilted. On the other hand, most of the scientists had made no attempt to really communicate, to talk in terms that would enable the public to see why the scientists were excited, or disappointed, or nervous about the outcomes. Lee thought a major, high-interest effort was needed to spread real understanding among both the media and the public.

Lee didn't let the matter rest with wishful thinking. Just one month after the second of the two Viking landings on Mars in 1976, he suggested to Sagan that they form a production company. They would call it Carl Sagan Productions. Sagan and Lee, who later returned to JPL as a key engineering leader in planetary missions, envisioned a company that would produce high-interest, cutting-edge TV programs and films about science. It would promote science by giving people a look inside, and conveying the mystery, the excitement, and the wonder of space exploration. Their company would be the Walt Disney of science and technology. Sagan was the creative portion of the team, while Lee would use his formidable project-management skills to run the production end of their projects. In fact, Lee is an extraordinary, multitalented individual, with a strong writing talent of his own, and the claim is sometimes made that, with Sagan, he coauthored *Cosmos: The Story of Cosmic Evolution, Science and Civilization*, the companion book to the TV series. However, his byline does not appear on the book. Later in his career, Lee would coauthor several science-fiction novels with Arthur C. Clarke, including the *Rama* series (1989–1993), and several other novels of his own.

Sagan agreed with Lee's visionary idea, but he took a little selling, as he was not a great admirer of TV and its less-than-elevating standard fare. In fact, in his speeches, he often made fun of TV's mediocre dialog and senseless sitcoms. So he was cautious. But he began to buy into the idea. He felt TV sponsors and programmers were underestimating the intelligence of the average viewer. He also hoped programming about real science would diminish the influence of pseudoscience, frauds, and hoaxes. Sagan was keenly aware that democracy requires a thoughtful electorate, not an electorate of fuzzy thinkers who are swept along on any bandwagon that comes by. People need to be ready and able to evaluate products: Are they safe? Do they work? They need to know how to question the motives behind government actions and understand the consequences. Who gains from this legislation? Is this a genuine proposal or just an election-year ploy? And they need to recognize the value of the scientific method of examining, hypothesizing, testing, and reaching conclusions based on evidence and reason.

So Sagan was primed when KCET-TV, a PBS station in Los Angeles, contacted him within a matter of days about doing a TV series on science that would parallel the popular series *Ascent of Man*, narrated by Jacob Bronowski. The proposed series would explore the human responsibility to explore the unknown. The concept immediately caught Sagan's interest, especially because he had quoted Bronowski in *The Dragons of Eden* and admired the integrity of the *Ascent of Man* series.

Sagan and Lee agreed to do a 13-episode program, to air in the fall of 1980, and set to work on it at once. Sagan devised the 13-week outline of the sexies, which he called *Man and the Cosmos*. (Druyan later convinced Sagan the title was sexist, so it became just *Cosmos*.) That was probably the simplest task they would do on the project. It would be a demanding, four-year process that involved attention to thousands of details, from conceptualizing shots and content to overseeing the creation of state-of-the-art (for the time) special effects and artwork. For the duration of the undertaking, Sagan, Druyan, and Sagan's secretary, Shirley Arden, moved to Los Angeles.

WILLS CLASH BUT LOVE GROWS

It wasn't destined to be easy. After conducting a search, Lee and Sagan hired British director Adrian Malone, who was the director and cowriter of the *Ascent of Man* series. The choice seemed ideal, as *Cosmos* sprang from Bronowski's series, which had won many awards, including an Emmy for best documentary film. And in the beginning things went fairly smoothly. However, a major conflict quickly came to light. Malone was used to having a free rein, and he was used to being in charge. But Carl wanted control. This central feature of his character would not, could not yield. He had too much on the line: his reputation both as a scientist and as a popularizer of science. It was too close to the core of his life mission for him to yield the reins. In any case, delegating was not in his nature. Malone, meanwhile, told Shirley Arden he planned to make a star of Sagan and endeavored to commandeer her help. The incident did not sit well with Arden, whose loyalty and allegiance to Sagan over the years she worked for him was unyielding.

Gentry Lee liked Malone, however. He saw the director as "voluble, fun, full of energy." Lee, who wryly claimed the software designers he had worked with had prepared him for Malone's egotistical side, was not put off by Malone's efforts to prove he was the most intelligent person on the set. A bit defensively, Malone often reminded Lee that he was from inner-city London, where he had learned to be the one that got to the top of the heap, a striver.

According to some members of the team, Sagan soon alienated Malone by an uncharacteristically unprofessional move. He missed the first production meeting, a meeting which was intended to set the tone for the entire project. Sagan was at the heart of everything. By his own design, he was the creative core, the fountain from which every detail flowed: the content, the mission, the writing, the form of the entire series. What was more, he was the star. But Ann Druyan and Carl Sagan were in love. Without thinking, they escaped together to Paris on the very day Malone had set aside for the production meeting. Malone was livid. And he was embarrassed. He had called together the entire production team, and the two most key players had not bothered to turn up. Without the two of them, he was stymied. Things were off to a bad start.

However, this gaffe was most likely caused by miscommunication and their working relationship could have healed. According to Druyan, the rift between the two men went deeper, caused by differences in values. In Malone, Sagan saw a sexist, racist, bigoted man with an annoying British style. Malone, meanwhile, saw himself faced with a controlling, exacting, egotistical star. He was used to being in control, and from his viewpoint, Sagan must have seemed to snatch his directorial powers away at the outset. Sagan liked to involve everyone in the process, holding "roundtable" discussions with the 40 crew members and seeking their opinions. To Malone, this was a waste of time. To Sagan, it was essential. He wanted the series to speak to everyone. "He wanted to be part of the social fabric," says Druyan. "He was writing this, he was conceiving this for the ages.... It was the accrued vision of his lifetime." Sagan was aiming for a tight production with fierce integrity—a presentation that would stand up 25 years later. And it does, adds Druyan. "I still get e-mails and letters every day from people who say that [*Cosmos*] touched them to their souls and made them want to be scientists or science teachers" (Druyan, 2004).

At the time, Ann and Carl were oblivious. In an interview with biographer Keay Davidson, Steve Soter (who, with Sagan and Druyan, was the third writer on the team for *Cosmos*) remarked, "This was the first time that Carl was really, thoroughly, and completely in love... He was glowing. His life was transformed by it. It was the most intense thing I've ever seen" (1999, p. 322). It was a love made to last for the rest of their lives.

Ann Druyan was tall, slender, and chic, with long, dark hair and the spark of intelligence in her eyes. Jewish (which pleased Carl's mother) and very bright (which probably pleased his father), Druyan received nearly everyone's vote; even Lynn Margulis thought Ann was very good for Sagan. In some ways a typical aspiring New York novelist and student of the time, she lived on a small inheritance from her grandfather and took on occasional small jobs. She said she was in and out of New York University, where she was an English major, and she freely admitted she was not a very good student. She later would say her learning only really began after she was out of school. Politically more liberal than Sagan, she hoped to raise his political consciousness—and was very likely to succeed. She was also an activist who had marched in many protests and been teargassed many times (Davidson, 1999, pp. 311–312). Her novel, *A Famous Broken Heart*, was published by a small publishing company in 1977, and now she was one of three writers holed up in Los Angeles working on a 13-week TV series—it looked like a great beginning.

In looking back on writing *Cosmos* with Carl and astronomer Steven Soter (whom Sagan had mentored at Cornell), Druyan recalls an idyllic process and the richness of the experience. By all other accounts, working on *Cosmos* was difficult, taxing, and nerve-racking. But Ann clearly derived joy from her work with both Carl and Steve. In their writing, all three were perfectionists, always searching for the right word to convey exactly the right nuance. Druyan would later say: "Here were two of the most brilliant people I've ever met, and two of the greatest teachers on the planet. Not only could you pose any question to them, but also you could count on an answer that would not only be scientifically impeccable, but would resonate with the wonder that the question and the answer implied" (Connell, 2002).

Druyan recalled working many all-nighters on *Cosmos* "and having a sense of just pure, unalloyed joy. There were moments of extreme frustration. With Carl there were no arguments from authority. You always had to build a case. The three of us would agonize over every single adjective, and every sequence, and everything that was there. And there were usually 25 iterations of virtually every script and every manuscript that we did together... But what a feast of ideas, of joy, of ecstatic communion" (Connell, 2002).

Yet on the set, the pressure brought out Sagan's abrasive side, and staffers would become offended. Off the set, Sagan insisted on keeping control and refused to delegate or grant authority. People couldn't do their jobs. Yet Sagan couldn't possibly do it all. He also insisted upon luxury hotel rooms, his favorite brand of chocolate milk, and iced tea prepared just so (the last two to help overcome the effects of achalasia). Sagan pushed himself, though, struggling for the right word, the right effect. Maybe, as Soter suggests, the few perks he requested were necessary to provide solace and calm so he could make it through the day.

Also, Carl's divorce from Linda Salzman was beginning to get messy. She was understandably hurt, and even Carl understood that. Now she seemed to be trying to even the score by costing him money and throwing lawyers at him.

Meanwhile, the schism with Malone did not heal. Sagan was generally ham-handed on the set, unaware that telling a crewmember how to do his or her job implied a disrespect for the crewmember's professionalism. For example, he might bluntly state the lighting wasn't right or a camera angle was ill chosen, without understanding how offensive his bluntness was, or his tone. Lee and the crew took to referring to this habit as Sagan's "Delphic oracle" mode; he seemed to think he could do or say no wrong, that he knew all there was to know on any topic, unlike everyone around him, who knew nothing about anything. Sagan was under extreme pressure, to be sure, but he did not respond to that burden with grace. Malone, in particular, understandably expected respect. He probably also expected to do some of the writing, as he had on Bronowski's series. He most certainly expected his talents to be used and his judgment to be accepted. And he expected to have control. Both grumbled about the other to Lee, whose take on their complaints was that they both thought they knew everything and both were wrong.

Finally, they were both so angry that neither one would speak to the other. They wrote notes or asked someone to deliver a message verbally. Yet Carl, always in control, never actually lost his temper. A project with so much complexity naturally comes with its tensions, but the set of *Cosmos* had more than the usual rash of problems. As with all projects of such magnitude, deadlines were tough to meet, and all-night sessions and insistence on completion repeatedly saved the day. The great success of *Cosmos* and its aesthetic and intellectual contributions certainly is a tribute to the ability of its creators to work well under the most difficult circumstances. In fact, *Cosmos* researcher Deane Rink later credited the protean efforts of Soter and Druyan in particular.

Arden helped Druyan and Sagan through this period with her loyal support and understanding. In February 1979, she sent a message of encouragement for the series from Ithaca, reminding them (in all caps for emphasis): "REMEMBER, IT IS FOR THE AGES—AND ESPECIALLY FOR THE CHILDREN."

And *Cosmos* did capture the sense of timeless wonder to which Arden alluded. One scene shows Sagan walking through the great library of Alexandria, long since burned to the ground. Sagan seemed to travel through time to the Egyptian city's glorious past and found the library as it once looked. In fact, he was filmed superimposed over the synchronized film of a miniature model (a trick done much more simply today with digital computers). In other scenes, we see Sagan on a precipice overlooking the Pacific Ocean; next to the Pyramids in Egypt; or whirring through space aboard a specially built spacecraft, getting a close-up view of Jupiter and catching glimpses of the "balloon animals" Sagan thought might live there, adrift amongst the planet's storm clouds.

PRESSURES OF LIFE

While working on *Cosmos*, Sagan continued with many of his regular responsibilities. He edited *Icarus*, as always. He kept his hand in with the Laboratory for Planetary Studies, and other duties at Cornell. He responded to requests from NASA to help lobby in Washington, D.C., for funds to launch future space missions.

But Carl also had personal burdens to bear. The divorce with Linda got uglier. She spent a lot of money on legal representation and used delaying tactics, certainly aware that Sagan hoped to marry Druyan. Linda felt Sagan had plenty of money, and he should part with some of it for her and Nick. Sagan said he would be happy to give money to Nick for computer lessons, for example, but he would deduct any amount that ended up being used for obtaining legal counsel. Nick was caught in between. He and his mother were living in Los Angeles at the time, so all this involved direct, personal contact. On the plus side, though, Nick, who was about eight or nine, spent time at the studio, watching his father work on *Cosmos*. He was fascinated and later pursued a career in television.

Carl's father, Sam Sagan, was ill with lung cancer, and Carl moved his mother and father to California so his father could obtain treatment at a hospital in Los Angeles. Carl visited often, despite the grueling *Cosmos* schedule, and worried about his mother, who knew she was losing the man with whom she had shared her entire life. When Sam died, Carl suffered a great loss, too, the loss of a father who was unreservedly proud of his son and had always encouraged and loved him. Sam died in 1979, just under a year before the first *Cosmos* program would air. It was an achievement Sam would have enjoyed and been especially proud of. Now Carl could no longer offer his father this special gift, nor enjoy his father's appreciation of it. Percolating deep inside Sagan was a story, a novel named *Contact* that, in part, would come out of this pain. It would be a story (and later a movie) about a SETI scientist named Ellie Arroway, who does what Sagan and his colleagues had never succeeded in doing: she makes contact with an extraterrestrial civilization. Here Carl's imagination takes off into the no-holds-barred world required of science fiction. Ellie visits the alien civilization (using blueprints for a spacecraft provided by the aliens from their distant home). When she arrives, she meets with a representative who has taken the shape of her dead father. The episode reveals Sagan's own grief and effort to deal with the eternal absence of a father he loves.

THE TURMOIL OF POPULARITY

When *Cosmos* finally aired in the fall of 1980, it became a runaway hit. It won an Emmy for excellence in television and the Peabody Award. The companion book, *Cosmos*, rose to the top of the *New York Times* bestseller list, remaining on the list for 70 weeks—close to a year and a half. The popularity was astounding and unparalleled, especially for a TV show on science. It remained the all-time most popular PBS show for a decade, not eclipsed until 1990, by Ken Burns's *Civil War*. Now, however, Sagan had a new problem—being known and recognized all over the world. *Cosmos* transported him from the well-known scientist people may have seen on Johnny Carson's late-night talk show to star status. Being that prominent, of course, also brought out the critics. Also, as a result of all this popularity, however, Sagan's books all began to make the bestseller lists, and he became an authentic pop star among the young.

That level of popularity brought out the critics, with reactions ranging from cheering bravos to grouchy bashing. Instead of rejoicing that suddenly children, students, and adults alike had a new and vital appreciation for science, some scientists became testy, jealous, and petty. Harvard astronomer Jay Pasachoff called the series "atrocious, embarrassing" (Davidson, 1999, p. 332). Others complained the pace was too slow. Still others thought Sagan's face filled up the screen too often (a directorial choice to which Sagan acquiesced, to his later chagrin).

Even Sagan's old friend Johnny Carson made fun of his eccentric pronunciation, the overly enunciated consonants, and sonorous cadences. The phrase "billions and billions" became a favorite punch line for an entire generation. Sagan finally reached the point where he would not utter the benighted word, because whenever he did audiences would titter. This audience reaction detracted from any serious discussion, but avoiding the word—for an astronomer—was no easy task, as indeed the vastness of the universe defies description without it.

But most people remember the wonder of *Cosmos*: the outstanding special effects done with great care (and great effort, in those predigital times), the verbal images and vivid examples (both verbal and visual), the excitement of visiting other planets and exploring space. Even when the details faded, the lasting impression was exactly what Sagan wanted it to be: an overpowering sense of awe at the enormity and complexity of the mysteries of the universe. Sagan's mother, Rachel, watched with pride, by then returned to her home in Florida, and Shirley Arden mailed her photocopies of the flood of fan letters that came in. An entire generation of scientists, science teachers, science writers, and many others were touched and motivated by *Cosmos* and for that we all owe an enormous debt of gratitude to Carl Sagan, Steve Soter, Ann Druyan, and all the others who worked so hard.

The 13-week PBS series may not be Carl Sagan's most important achievement scientifically speaking, but in many ways *Cosmos* was his crowning accomplishment. Its enormous influence for the cause of science places Sagan among those few who have provided a new way of seeing things and changed people's lives. *Cosmos* reached some half-a-billion people in 60 countries with a message about science that had never before been so effectively delivered. It was his gift to the world.

As Gentry Lee put it, "Carl probably touched more hearts and minds than any scientist in history. I would guess that more people decided to pursue scientific careers because of him than any person in history. For all these things the world should be eternally grateful" (Davidson, 1999, p. 339).

Chapter 8

POLITICAL ADVOCACY AND ANNIE'S INFLUENCE

Ann and Carl were a match. They complemented each other in all the ways that just never had happened in Carl's previous relationships. This magic was not really magic; it was just the right combination. It resulted in part from the fact that Sagan recognized the youthful mistakes he had made before. But also, these two were partners. They worked together on writing projects and their creativity flowed easily from one to the other. All Sagan's writing began to show Druyan's influence, and they both found joy in their work together. She would later say of her work with Carl, "It was a feast for me. It was the greatest experience of my whole life" (Connell, 2002).

In May 1981, Sagan's divorce became final, and Carl and Annie married on June 1 of that year. It was the fourth anniversary of the day they confessed their love for each other. The celebration took place at the Hotel Bel-Air in Los Angeles. Carl's sister Cari was there, Ann's brother was there, and a host of friends who one by one got up and toasted Carl's new, gentler personality, tamed by Annie. Each toast brought down the house, with Carl laughing perhaps the hardest of all. (Privately, though, Ann came to Sagan's defense, asserting that she never knew Carl to say or do a mean thing and that she had learned a good deal from Carl about being a human being.)

In many ways both friends and some enemies saw Sagan's marriage to Ann as a new chapter; she seemed to mellow him. He now appeared more relaxed, less elitist, and more ready to listen to the views of others.

Druyan, who freely points out she was always farther left in the political continuum than Carl, also had a lot to do with awakening a sense of activism in him. He felt a deep concern about the expanding arsenals of nuclear weapons. That haunting question raised years ago—What is the value of *L?*—continued to beleaguer him—*L*, the length of time that a communicating civilization may survive. Do technological societies always self-destruct? Planet Earth continued to teeter at the edge of a potentially fatal precipice. Political factions—headed up by the United States and the USSR—playing deadly games of chicken for the sake of power.

Sagan's name was by now a household word. He had become a celebrity, not joyfully in every way. The lack of privacy that came with his fame sometimes bothered him. But a kind of power, an ability to influence, also comes with fame. Sagan had gained the respect and the attention of millions of people. Now was the time, he decided, to use that influence to combat a problem that truly frightened him when little else did.

NUCLEAR WINTER

Ever since Sagan first encountered Drake and his equation, Sagan realized that the silence from SETI searches were not necessarily caused by inadequate instruments, vast distances, the need for better search strategies, or the small amount of time actually devoted to the search. Another explanation for the silence might be that the value of *L*, the length of time a technologically sophisticated civilization will survive, is so low. Maybe, as long as the universe has existed, very few or no civilizations have perpetuated their existence beyond a few short years or decades. Even a hundred years would be brief compared to the timescales of solar systems and galaxies. The dinosaurs, which have been extinct for about 65 million years, ruled Earth's animal kingdom for 165 million years. Humans, by contrast, have been around for only a tiny fraction of that time. But dinosaurs did not have the power to annihilate themselves with nuclear weapons or other agents of mass destruction. A technologically sophisticated civilization such as our own does have that power and we have been sitting precariously on the precipice of nuclear holocaust and global warfare ever since it was obtained. How long can such a civilization hope to avert this disaster?

Since the United States detonated two nuclear weapons at the end of World War II, people have been fully aware of the effects of such a blast on people, the environment, and structures. The damage inflicted is far more devastating than that of any previous weaponry. But in the early 1980s, Sagan and other scientists became concerned about effects on the climate caused by widespread ash and fire thrown up into the atmosphere, where smoke and ash would remain suspended for a year or more, effectively blocking out sunlight, annihilating plant life and other forms of life, and causing a global ice age. This was a phenomenon that came to be known as "nuclear winter."

Sagan and other authors introduced this term in an article published in *Science* in December 1983. The article postulates that after nuclear explosions take place, smoke particles from subsequent fires could spread globally, absorbing the sunlight and darkening the sky. Surface temperature could be lowered by as much as 5 degrees Centigrade. The study showed that a drop of even just 1 degree Centigrade could destroy many species, including humans.

Sagan's study of the Earth's atmosphere led him to formulate the idea of nuclear winter with American scientists Paul and Anne Ehrlich in the 1980s. Sagan and the Ehrlichs theorized that the dust and ash thrown into the atmosphere by the explosions of nuclear bombs and the ensuing fires, might be so thick and widespread that it would block the sunlight for months or years. The damage that a nuclear winter would cause to crops and ecosystems would be at least as devastating as the nuclear explosions. The idea of nuclear winter was met with much controversy, and scientists continue to debate the theory.

This line of reasoning sprang from the findings of the father-and-son team Luis and Walter Alvarez. While digging in a layer of rock that was laid down 65 million years ago, Walter Alvarez, a geologist, found a substance he couldn't identify. He called his father Luis, a Nobel laureate in physics. Luis Alvarez examined the material and saw that it was radioactive substance called "iridium," which is common in space, but not on Earth's surface. This evidence added to a body of evidence built up in the early 1980s indicating that the dinosaurs were destroyed by a cataclysmic collision of an asteroid into the Yucatan Peninsula in Mexico. Ash and smoke from fires ignited by the collision dispersed throughout the atmosphere, and most of the living organisms on Earth were destroyed. Since the first findings, much more confirming evidence has been found, and this explanation of the long-mysterious death of the dinosaurs is generally accepted today.

Sagan and his colleagues saw a clear parallel between that scenario and the destruction that humans could wreak with nuclear weapons. Their estimates found that the use of even just half of the combined U.S. and Soviet weapon stockpiles would block out sunlight for months, creating a subfreezing climate. They also thought the ozone layer might become damaged, which would cause further problems due to UV radiation from the Sun. If this situation were to continue for long, human civilization might be wiped out. These findings received further support from a report released by the U.S. National Research Council in December 1984, but in 1985 the U.S. Department of Defense acknowledged the validity of this research, yet announced that no change would be made in U.S. defense policies.

Sagan began lobbying against nuclear weapons and wrote articles about nuclear winter. At the same time he became a vocal opponent of the U.S. Star Wars defense strategy, a space-borne antimissile defense.

Druyan and Sagan also often attended rallies and protests, especially against nuclear weapons. In 1987, for example, they were spotted protesting at the Nevada test site, where the federal government conducts underground nuclear explosions, including those for the Strategic Defense Initiative (Star Wars) program. Protestors blocked the entrance, briefly preventing more than eight thousand workers from entering the site. In some instances, Druyan and Sagan were arrested, along with other protestors.

Sagan continued to battle against nuclear testing and other misapplications of science, using his popularity to wage war for his political beliefs. This tactic began to change the playing field in which he operated, winning him many new admirers but also many new enemies. For those who knew him best, though, this move signaled that Carl was no longer grandstanding as he once did, but was now intent on using all his talents to speak for his convictions, not just to gain fame or further his career.

Since the dissolution of the Soviet Union and the end of the Cold War, the issue of nuclear winter has lost some of its sense of timeliness, despite the proliferation of weapons of mass destruction to many other countries in some ways making the likelihood of a nuclear exchange just that much greater.

SCIENCE AS A PUBLIC CAUSE

Sagan embarked upon a period of outspoken involvement in issues involving science and public policy and became an active and influential supporter of the Committee for the Scientific Investigation of Claims of the Paranormal (CSICOP). As a fellow of CSICOP, he spoke out clearly and frankly against pseudoscience, supporting the no-nonsense views of author Martin Gardner (whose book *Fads and Fallacies in the Name of Science* taught Carl early in life to question his own views about UFOs).

Sagan was a dedicated and tireless public educator and would explain carefully to listeners and readers how science works. He liked to point out that science rarely claims any piece of knowledge is certain, whereas pseudoscientists are more likely to claim they have the entire and final answers. This is because science, by its nature, is self-correcting. Science takes the bigger view, recognizing that facts may seem to fit one way of seeing the universe (or some portion of it) today, but tomorrow new information may require adjustments—or sometimes revolutionary change—to the observations we have made and in the way we see the universe.

PROMOTING PLANETARY SCIENCE AND SETI

Faced with the dichotomy of high public interest in space exploration and low political support, planetologist Bruce Murray and astronautics engineer Louis Friedman joined with Sagan to found the Planetary Society in 1980. The group's purpose was twofold: to advance exploration of the solar system and to encourage and support SETI. Mailings went out bearing an open letter from *Star Trek* author Gene Roddenberry encouraging *Star Trek* fans to join up. From that jumpstart, the Planetary Society rapidly became the largest space interest-group in the world, with membership swelling to one hundred thousand, and members in 140 countries. Thanks to Sagan's enhanced lobbying power, given the strength of his new constituency, Congress paid more attention to Sagan's expert testimony on behalf of the space program. The Planetary Society continues to take activist stances by campaigning for exploratory missions, such as a human presence on Mars, and missions to Pluto (never visited so far) and Europa (which may harbor life forms in the ocean beneath its icy crust). The society's logo, created by space-artist Jon Lomberg for publication in Sagan's book *The Cosmic Connection*, represents a Renaissance sailing ship morphing into a sketch of the Voyager spacecraft—linking the explorers who sailed the uncharted seas of times long past with present and future explorers of the cosmos.

By 1983, the society was funding its own SETI project using a small radio telescope at Harvard, and because of his interest in SETI, actor Paul Newman donated a check for $10,000. The new SETI search was conducted by Harvard University physicist and SETI astronomer Paul Horowitz, whose observations among the galaxies fascinated Sagan. Since the days of Frank Drake's Project Ozma, the technology and the equipment used by SETI astronomers has come a long way. "We are literally the first generation that could communicate with an extraterrestrial civilization," Horowitz remarked in a PBS Online *Nova* interview. "The evolution of radio astronomy, the large antennas, the receivers, the computing hardware is unique to our generation and has not existed before. We... are the first generation that could establish contact.... Contact, receipt of a message would be the first bridge across four billion years of independent life . . . and evolution. It would be the end, in a very deep sense, of our earth's cultural isolation. It would be, I think without doubt, the greatest discovery in the history of mankind" (Horowitz, 1996). Despite his experience at Arecibo, Sagan never lost hope of finding a signal—an undeniable sign of intelligent life. Like Sagan, Horowitz is optimistic, pointing out that only a few dozen searches have ever really been done since Drake's effort in 1960. Earth, he says, has spent little time listening, and the universe is big, and the radio channels are many. Only since the mid-1990s have multimillion-channel receivers been available, like the one used in the Planetary Society search. During his years with the Megachannel Extraterrestrial Assay (META) project, Horowitz has heard several signals, but they always appear once and then disappear. No signal that comes from space has ever been detected more than once. Why?

As has sometimes been shown to be the case, the origin of such a signal may not be the stars at all, but an overlooked signal from Earth, perhaps a military satellite or other dark object. Or it may be a case of scintillation, the twinkling we see in starlight, caused by interferences caused by ionized matter between the observer and the observed object. Sometimes this kind of phenomenon also affects radio signals. Radio signals from clear across the country will sometimes bounce thousands of miles out of their territory, and that's when a truck driver on the way across the Rockies to Missoula, Montana, may suddenly pick up a radio program broadcast from Miami, Florida. So maybe, Sagan postulated, a sort of scintillation occurs that enables a very distant radio signal to come through clearly for brief moments and then the signal is gone. Perhaps some of Horowitz's brief signals, Sagan thought, might be real messages from another civilization, interrupted by ionized material and lost.

Sagan's interest in SETI was a lifelong commitment, despite the lack of any encouraging discovery. SETI scientists just keep trying to build a better signal trap. The community of SETI scientists appreciated Sagan's ground-floor presence in SETI research and his steady moral support and contributions. By the late 1970s, both California NASA research centers, JPL and Ames Research Center, had their own SETI programs, which were coordinated to develop a dual program with JPL doing full-sky surveys and Ames using a targeted search that focused on 1,000 stars resembling the Sun. The SETI teams put in a decade of planning, and just after their search was finally funded by NASA in 1988 and the searches began four years later, Congress ended all SETI funding in 1993. Through private funding, the large-scale project has continued in the form of the SETI Institute in Mountain View, California, and its Project Phoenix (which conducted searches much like those done by protagonist Ellie Arroway in Sagan's book *Contact*, and Ellie in turn resembles the Institute's Center for SETI Research director Jill Tarter). Using Arecibo and other large antennas around the world, SETI scientists have a systematic, thorough plan in place, and with this worldwide approach—one supported by Sagan—they hope to achieve the so-far elusive goal of finding a signal that will reveal the presence of intelligent life on another world, somewhere many light years away.

In 1997, the SETI Institute established the permanently endowed Carl Sagan Chair for the Study of Life in the Universe. Building a privately financed SETI program and establishing this endowment was an important part of how Sagan used the money and fame that came out of *Cosmos* and *Contact* (for which he received a $2 million advance from Simon & Schuster). The chair is currently held by Christopher Chyba, who, in addition to being a Marshall Scholar at the University of Cambridge in England and a MacArthur Fellow for his work in both astrobiology and international security, is also a former student of Sagan's at Cornell.

FATHER AND GRANDFATHER

Encouraged by Druyan, Carl renewed relationships with his estranged children. Ann would say to him firmly, "Carl, *be* a father." And she gave birth in November 1982 to Carl's only daughter, Alexandra Rachel Druyan Sagan—Sasha for short—and in 1991, to a son, Samuel Democritus Druyan Sagan. Carl seemed to settle into the joys of being a father like never before. He was soon to discover another pleasure of family life. In 1984, Dorion had a son, Tonio, whom Carl loved dearly, and being a grandfather became an important part of his life.

He also soothed some wounded feelings. It had been 20 years since he and his first wife had parted ways. Now, he met with Lynn Margulis, took her out, and apologized. He now realized how unfair he had been about housework, about his views of what his responsibilities were in a marriage. Margulis feels sure Carl's understanding that he owed an apology came from Annie, that she "made him do it," for Lynn had already sensed there was a bond between her and Annie, an unspoken understanding. As a writing and creative partner, Ann received a respect from Carl that no other person ever had. She could give him the attention he needed because she was part of him. And also, maybe fame had finally cured his need for the rapt attention of an audience of ten thousand.

Lynn also told both Carl and Ann that she was sorry for hurtful things she had done as well. With Druyan's encouragement, he wrote his only work of fiction, *Contact*, published in 1985, a bestseller about using science to search for intelligent life on other planets and what might happen when scientists do make contact with intelligent beings from outer space.

REJECTED AND HONORED

Fame both helped and hurt Sagan as a scientist. He became America's best-known science writer and popularizer; yet some critics question his importance as a scientist. Sagan received a deeply disappointing blow when the National Academy of Sciences rejected his application for membership in 1992. The Academy reserves it membership to what some members call an "old boys club." Sagan's former wife, Lynn Margulis, who was elected a member of the Academy in 1983, was outraged at the behavior of her colleagues, and she wrote Sagan a letter of praise and apology. Ironically, two years later the Academy awarded Sagan its Public Welfare medal, described as the "Highest award of the academy," for "his distinguished contributions to the application of science to the public welfare." The text of the honor read: "Carl Sagan has been enormously successful in communicating the wonder and importance of science. His ability to capture the imagination of millions and to explain difficult concepts in understandable terms is a magnificent achievement."

Chapter 9

FINAL YEARS

Even before writing *Contact* and its publication in 1985, Sagan had entertained the idea of making the story into a film. He had already spent time in Movieland, consulting with director Stanley Kubrick and author Arthur C. Clarke on making the film version of Clarke's novel *2001*. He had even once talked with Francis Ford Coppola about possibly making a film about a search for extraterrestrial intelligence (a fact that prompted Coppola to sue Sagan's estate after his death).

Producer Lynda Obst (who had been at that first meeting of Druyan and Sagan at Nora Ephron's house) eagerly jumped at the idea of making a film with them. Educated in philosophy, especially the philosophy of science, at Columbia University, Obst was a natural as producer for this film. *Contact* is a gentle, cerebral story in which Sagan clearly tried to embrace the disparate parts of his being, the objective, strongly rational, left-hemisphere self and the intuitive self who wished he could believe he would one day see his father again. The lead role is SETI scientist Ellie Arroway (whose name evokes the unwavering human-rights activist Eleanor Roosevelt, combined with the sense of an homage to the sharply rational and naturalistic thinking of French philosopher François Arouet [pronounced "arroway"], better known by his pen name, "Voltaire"). Arroway is an unexpected character in a Hollywood film. She is a serious scientist (not unheard of, but Helen Hunt's role as a meteorologist in *Twister*, by contrast, is mostly that of an action hero on the run), a woman, and an atheist. The latter characteristic is practically unheard of among Hollywood heroes and heroines and risked rejection even before the box office opened, but Sagan considered honesty about this issue, especially, to be important. In *Contact* the movie, the opening scene sets much of the tone. A young girl (Ellie) looks at the sky through a backyard telescope, and the movie camera sweeps through stars and galaxies to enter the telescope and encounter her eye. The greatness of the universe has entered her life. At the same time her father, who has raised her alone and taught self-sufficiency and many insights about how the world works, is dying, slipping away from her. It is a moment of great discovery and great loss. The adult Ellie, played by Jodie Foster, embodies intelligence, cool objectivity, and honesty.

Obst, Druyan, and Sagan spent hours writing the movie script together, recording their conversations, and hashing out every detail. Some readers of the novel felt that the movie copped out, portraying Ellie, in the end, as weak, not truly seeking knowledge and contact for objective reasons, not searching for scientific reasons, but instead just looking for love (as every woman does, the ending seems to imply in an unfortunately sexist moment). The movie diverges and shows the effects of input from many sides. Indeed, movies are notorious for failing to reflect the original intentions, which usually wind up on the cutting-room floor. The project had been through several hands already, and one early studio mogul had told Obst that Ellie basically had to be a neurotic woman; no other interpretation could explain why she was seeking scientific truth. Given Sagan's obvious appreciation for strong women and feminine rights, it is doubtful he intentionally wrote a sexist ending. He probably piled too many messages on the shoulders of Ellie Arroway—representing women in science, atheism, and emotions that Sagan himself needed to work through. At the end of the novel, Ellie says, "I wish I had a baby," and the team working on the film seriously considered having her become pregnant at the end of the story. But that would have muddied the story even further, and luckily they saw that.

This film was Sagan's pièce de résistance, the last creative task he would ever work on. Much about the movie talks directly to people who see adventure in discovering the universe, the wonders of galaxies, the well-kept secrets of the planets, the mysteries of the moons. It also speaks to those who have never thought of the stars and planets that way. And it provides a model that says to young women, "You see, you can have this joy, this satisfaction, this special journey. Because science is not just for boys."

CELEBRATING LIFE

On Carl's 60th birthday, November 9, 1994, the Cornell Astronomy Department and many friends threw a huge birthday party for him in Ithaca. It was a nonstop evening of luminous speeches, with friends, colleagues, and loved ones taking turns to honor him with praises. An undergraduate student told how a single meeting with Carl resulted in a decision to major in science. A graduate student told of how he had realigned his thesis to focus on planetary atmospheres as a result of a discussion with Carl about the exciting developments in that field.

The chair of the astronomy department read a letter from a student in Africa who had organized an astronomy club in his village, Niamey, Niger, after reading *Cosmos*. Topping that, the department had flown the student to Ithaca for the occasion, and he was personally introduced to Sagan. Letters from top officials from several countries also were read. Each of these individuals represented hundreds of millions of others touched by Carl Sagan through his words and his passion for science, and his excitement about the process of scientific inquiry and discovery.

Many former colleagues and old friends attended, including Kip Thorne, James Randi, Frank Drake, and Philip Morrison. Some could not come—so their letters—from the likes of British science-fiction writer and visionary Arthur C. Clarke and then–Vice President Al Gore—were read aloud.

Most of Sagan's family was there, which touched Carl. All of his children attended, as well as Annie Druyan and Lynn Margulis (though not Linda Salzman, who remained distant).

The top of the evening came with the announcement that Eleanor Helin, an expert in near-Earth bodies and a leading discoverer of asteroids, had named her discovery, asteroid 4970, "Asteroid Druyan," after Annie Druyan. Her asteroid's orbit resonates in an eternal dance with another (asteroid 2709 Sagan), named after Carl. It was a highlight of the evening—the perfect birthday present, ensuring they would always be together, in a sense, in the cosmos.

JUST A BRUISE

Toward the end of 1994, Ann noticed that a large bruise on Carl's arm had been there for a while. Something about it didn't seem normal, so she convinced him he should see a doctor. He did and had the requested tests done and then went on with his life, which in this case meant a trip to Houston. While he was gone, his doctor called, requesting a retest. The results must be wrong, the doctor said. He was even more convinced the results were wrong when Annie told him Carl was on a trip. No one as sick as these results indicated could be up to traveling.

But the results were not wrong. In late 1994, Carl was diagnosed with Myelodysplasia, a bone-marrow disease sometimes called "smoldering leukemia." Sagan was told his chances were slim, but optimistic, as always, he continued working and kept on popularizing science.

THE VALUE OF MISTAKES

One of Sagan's projects during his battle for health was a book called *The Demon-Haunted World*. Its subtitle evoked his purpose, an evocation of *Science as a Candle in the Dark*. Written in a colloquial, approachable style, the same style he used for his frequent articles in the Sunday newspaper supplement *Parade*, this book was a collection of essays intended for everyone and anyone, and it captured his most warm and generous understanding of why people hang on to irrational beliefs. In 25 brief essays, he broached a wide variety of topics, from UFOs and why they are *not* here to the case against antiscience to thinking critically about government as a patriotic act.

In this book Sagan did something most scientists, indeed, most people, never do. He published a list of mistakes he had made. Explaining his objectives for including such a list, Sagan wrote: "It might play an instructive role in illuminating and demythologizing the process of science and in enlightening younger scientists." While he had been right, for example, about Venus's high temperatures and the greenhouse effect, spacecraft visits and other sources had proven him wrong on several counts. "At a time when no spacecraft had been to Venus, I thought at first that the atmospheric pressure was several times that on Earth, rather than many tens of times. I thought the clouds of Venus were made mainly of water, when they turn out to be only 25 percent water" (Sagan, 1996, pp. 256–257). The list goes on. Although one more book would go to press after his death (*Billions and Billions*), *World* would be the last book Sagan would see published.

Sagan's trumpet call for the case of reason remained strong. He warned against the highly combustible mix of power and ignorance that currently exists in our global society. More than ever before, he points out in his essay "Science and Hope," virtually all aspects of our society depend upon science and technology—"transportation, communications, and all other industries; agriculture, medicine, education, protecting the environment; and even the key democratic institution of voting." Yet few people understand (or, we could say, even try to understand) science or technology. Whenever emotional issues come up—insecurity, fear, prejudice, helplessness—as Sagan points out, "The candle flame [of science and reason] gutters. Its little pool of light trebles. Darkness gathers. The demons begin to stir." Historically, in such circumstances, witch hunts thrive, people are burned at the stake or otherwise executed for no cause, ordinary citizens are imprisoned without bail or explanation, genocides sweep across countries. Today is no different in these respects than times past. But we do have an alternative. Science, as Sagan points out—with its success in explaining how the world really works—can help us to become objective, to solve problems, and to overcome the real causes of disease and natural destructive forces (1996, pp. 26–27).

HOPE FOR RECOVERY

Sagan's doctor held out one hope. If a bone-marrow donor could be found, perhaps the progress of the disease might be slowed substantially. Carl's sister Cari stepped forward immediately, and more than once, renewing their childhood bond, but the procedure failed to save Carl's life. During his last two years, Sagan lived a nightmare of pain and physical deterioration, but despite these deterrents, he continued to write and gave a few important public speeches and appearances.

His face became gaunt and drawn, his hair disappeared, his eyes appeared tired. For a man just over 60, he looked shockingly old and debilitated, but Carl remained steadily optimistic. Recent news reports from NASA brightened his hope for discovery of some evidence of ancient life on Mars. The motion picture starring Jodie Foster as Ellie Arroway was underway based on his novel, *Contact*. There were so many things to live for.

In November 1996, the family gathered for Thanksgiving dinner. Carl's sister Cari was there, and all his children—the adult sons, Dorion, Jeremy, and Nick; his grandson Tonio (Dorion's son); and the younger children, Sasha and Sam. It was a fine reunion.

A few weeks later, Sagan flew to San Francisco to give a couple of lectures. Seeing he was exhausted that evening, Druyan called the cancer center where he had been treated in Seattle. It would turn out to be their last trip to the center. The X-ray revealed pneumonia.

"This is a deathwatch," Carl told Ann. "I'm going to die."

She resisted with that characteristic optimism she shared with her husband.

He replied, "Well, we'll see who's right about this one."

Finally, on the morning of December 20, 1996, Carl Sagan died. In a postscript to her husband's last book, published in 1997, Ann Druyan shared her courageous, unflinching thoughts, "As we looked into each other's eyes, it was with a shared conviction that our wondrous life together was ending forever" (p. 271).

Sagan and Druyan's romantic private story may capture the hearts of their admirers, but the honesty and integrity that filled their lives also win admirers' deep respect. We often measure the stature of a public figure by the prestige and number of medals and honors granted. We rush to grant them as life begins to fade, and in Sagan's lifetime, as one might expect, he was showered with awards for his work and contributions, and honored by election to key posts in his field. For his work, Sagan received the NASA Medals for Exceptional Scientific Achievement and for Distinguished Public Service twice, as well as the NASA Apollo Achievement Award. He served as Chair of the Division of Planetary Sciences of the American Astronomical Society, as President of the Planetology Section of the American Geophysical Union, and as Chair of the Astronomy Section of the American Association for the Advancement of Science. Even the skies now echo his name with the orbit of Asteroid 2709 Sagan. Sagan also received the John F. Kennedy Astronautics Award of the American Astronautical Society, the Explorers Club 75th Anniversary Award, the Konstantin Tsiolkovsky Medal of the Soviet Cosmonauts Federation, and the Masursky Award of the American Astronomical Society, which recognized him for "his extraordinary contributions to the development of planetary science" including "seminal contributions to the study of planetary atmospheres, planetary surfaces, the history of the Earth, and exobiology," noting the extraordinary number of today's productive planetary scientists who got their start under Carl Sagan's influence either as students or associates.

After Sagan's death, a great outpouring occurred, including thoughts, reminiscences, and special memorials. Carl Sagan was a great friend of reason, and he is sorely missed.

On July 4, 1997, a small NASA spacecraft literally bounced to the surface of Mars. *Pathfinder*, part of NASA's renewed emphasis on planetary missions in the late 1990s and early twenty-first century, was inexpensively produced, and its mission went flawlessly. After righting itself on the Martian surface, its rounded surface opened up like the petals of a flower, revealing the rover *Sojourner*, a small robot lab assistant. About the size of a microwave oven, *Sojourner* went from rock to rock, scraping samples, testing, exploring. Carl Sagan would have loved being at Mission Control in JPL. In a real way he *was* there in the minds of many of the mission scientists for whom he had been an inspiration. They named the landing craft the Sagan Memorial Station in his honor.

Contact, the film, made its debut on July 1, 1997 to great popular success. As the film ended, the words "For Carl" flashed upon the screen and brought a tear to the eyes of those who noticed it and understood.

The NASA news about possible evidence of Martian life remains controversial, but the movie *Contact* won rave reviews from critics and was a popular hit. And Carl Sagan leaves a brilliant legacy behind—a bright and shining light that continues to show the way to understanding science, its methods, its strengths, and the excitement of its discoveries. He has already galvanized an entire generation with the thrill of a universe filled with billions of stars and countless wonders and unsolved scientific mysteries far exceeding human imagination. And he is certain to do the same for many more generations to come.

In the words of science writer Shawn Carlson, "The ultimate measure of a great life lived is the impact one has had on other lives. And by this accounting, Carl Sagan was a great man indeed. His life and works will ultimately contribute to far more good in the world than even he could have known." His was a life that touched many others. Leon M. Lederman, director emeritus of the Fermi National Accelerator Laboratory and a 1988 Nobel laureate in physics wrote: "By his own passion for planetary science, by his active concerns in national science policy, and by his unique and heroic efforts in public understanding, Carl Sagan has set new standards for the conduct of scientists" (Dawkins et al., 1997, p. 12).

And he went on setting those standards right up to the end. NASA administrator Daniel Goldin likes to tell how one day in late 1996, Sagan stopped by to see him. He talked about the future of space explorations, visions he had for missions and projects. The discussion was lively, and they continued talking over dinner. Two weeks later he was gone, leaving behind his many visions and ideas. Goldin later remarked with obvious appreciation, "He was talking with intensity—a man on his deathbed. This is the Carl Sagan I love, a man so full of hope and optimism that he never gave up. He never gave up" (Broad, 1998).

To that British evolutionary biologist Richard Dawkins adds, "It is hard to think of anyone whom our planet can so ill afford to lose" (Dawkins et al., 1997, p.6).

Chapter 10

A BRIGHT AND SHINING LIGHT

On January 3, 2004, a rover called *Spirit* landed at Gusev Crater on Mars. It was the firsts of two rovers to land on the Martian surface as part of NASA's Mars Exploration mission. Three weeks later the second lander, *Opportunity*, landed at Meridiani Planum. That evening, as millions of television viewers watched, scientists and technicians rejoiced with jubilant yelps and excited hi-fives on the crowded floor of Mission Control. That evening both at mission control and at home watching, many became, once again, sadly aware of Carl Sagan's absence. But the absence was physical only. Carl's dreams, ideas, and ambitions had touched many of the scientists who so excitedly exchanged congratulations that evening. Many television viewers, as well, still remembered vividly the unwavering enthusiasm for science and its wonders he had displayed on television news and talk shows. To those old enough to remember the excitement of the *Viking* landing nearly thirty years earlier, time seemed suddenly to condense and collapse in upon itself. Memories welled up of Carl's love and fascination with Mars and joined with recollections of his continuous search for signs of other life in the universe to unite both the past and the present into one wondrous quest for the future making small all the trials and tribulations of everyday life. Such was the power of Carl's vision.

"In light of his marvelous communication skills, lots of people lose sight of how he was also a hell of a physicist," remembered Steven Squyres, who had taken courses under Carl at Cornell. Squyres was in a position to know. In the excitement as *Opportunity* began sending its stunning first pictures back to Earth, Squyres, as chief scientist for the rovers, was the point man for the science-team interviews (Broad, 1998).

"Holy smokes!" Squyres said as the first pictures from *Opportunity* came in. "I am flabbergasted. I am astonished. I am blown away!" (Chang, 2004).

The spontaneous enthusiasm of Carl Sagan was still alive at NASA. One can almost hear Carl saying those same words and imagine his face lit up in almost boyish excitement.

As Squyres noted, though, Carl was also "a hell of a physicist," a statement readily confirmed by Sagan's some five hundred scientific papers and contributions, but a fact sometimes overlooked in favor of the exuberance of his popular writing and speaking style. It was as a scientist that Sagan could speak so lovingly and enthusiastically about science, that he could evoke so knowingly the mysteries of the universe and the achievements of humankind in trying to unravel those mysteries. In an age when pseudomysteries could be cheaply devoured in books, television shows, and the movies; in an age when thousands of Internet sites titillated viewers' imaginations with wild tales of UFO abductions, eight-foot-high creatures roaming the backwoods of the American northwest, and mysterious faces carved into the rocks on Mars, Carl managed to elicit wonder without sacrificing truth.

As Ann Druyan knowingly observed in an interview with *Astrobiology at NASA* executive producer Kathleen Connell, "What Carl did was he reunified skepticism with wonder—and never one at the expense of the other, but always in equal parts. I think that what we all long for is something that could raise goose bumps, could make you feel something in your heart without requiring that you lie to yourself" (Connell, 2002).

For those of us who follow, Carl Sagan has left a bright-and-shining light, pointing out the too-little-used paths of reasoned thought and action, the too-seldom-traveled roads of scientific inquiry, and the mighty wonders of the cosmos.

Gifted with a marvelously creative mind tempered by a love for the self-discipline of science and the scientific search for truth, whether in the vastness of the Universe or in the dark corners of the human imagination, Carl Sagan not only represented the best of science in our time, but created a template for the science of the future. As the twenty-first century unfolds with its promises and its threats, its hopes and its fears, we forget Carl Sagan's legacy at our peril, but remember it with our most ambitious dreams.

SELECTED BIBLIOGRAPHY

BOOKS AND ARTICLES BY CARL SAGAN

Sagan, Carl. (1961). *Organic Matter and the Moon*. Panel on Extra-Terrestrial Life for the Armed Forces-NRC Committee on Bio-astronautics. Washington, DC: National Academy of Sciences-National Research Council.

Sagan, Carl. (1961, March 24). The Planet Venus. *Science*, 849.

Sagan, Carl. (1963). Direct Contact among Galactic Civilizations by Relativistic Interstellar Spaceflight. *Planetary and Space Science*, *11*, 485–98.

Sagan, Carl. (1970). *Planetary Exploration*. Eugene, OR: State System of Higher Education.

Sagan, Carl (Narrator). (1972). *Discovering Mars* (Sound recording). Washington, DC: American Association for the Advancement of Science.

Sagan, Carl (Ed.). (1973). *Communication with Extraterrestrial Intelligence (CETI)*. Cambridge: MIT Press.

Sagan, Carl. (1973). *The Cosmic Connection: An Extraterrestrial Perspective*. Produced by Jerome Agel. Garden City, NY: Anchor Books.

Sagan, Carl. (1977). *The Dragons of Eden: Speculations on the Evolution of Human Intelligence*. New York: Random House.

Sagan, Carl. (1979). *Broca's Brain: Reflections on the Romance of Science*. New York: Random House.

Sagan, Carl. (1980). *Cosmos*. New York: Random House.

Sagan, Carl. (1985). *Contact: A Novel*. New York: Simon and Schuster.

Sagan, Carl. (1994). *Pale Blue Dot: A Vision of the Human Future in Space*. New York: Random House.

Sagan, Carl. (1995). *Cosmos*. Avenel, NJ: Wings Books.

Sagan, Carl. (1996). *The Demon-Haunted World: Science as a Candle in the Dark*. New York: Random House.

Sagan, Carl. (1997). *Billions and Billions: Thoughts on Life and Death at the Brink of the Millennium*. New York: Random House.

Sagan, Carl. (2002). *Cosmos*. New York: Random House.

Sagan, Carl, & Druyan, Ann. (1992). *Shadows of Forgotten Ancestors: A Search for Who We Are*. New York: Random House.

Sagan, Carl, & Druyan, Ann. (1997). *Comet*. New York: Ballantine Books.

Sagan, Carl, et al. (1978). *Murmurs of Earth: The Voyager Interstellar Record*. New York: Random House.

Sagan, Carl, et al. (2000). *Carl Sagan's Cosmic Connection: An Extraterrestrial Perspective*. Produced by Jerome Agel; new contributions by Freeman Dyson, Ann Druyan, & David Morrison. New York: Cambridge University Press.

Sagan, Carl, Leonard, Jonathan Norton, & the editors of *Life*. (1966). *Planets*. New York: Time, Inc.

Sagan, Carl, Owen, Tobias C., Smith, Harlan J., & Dordrecht, Reidel (Eds.) (1971). *Planetary Atmospheres*. New York: Springer-Verlag.

Sagan, Carl, & Page, Thornton (Eds.). (1974). *UFOs: A Scientific Debate*. New York: Norton. (Originally published by Cornell University Press in 1972)

Sagan, Carl, & Turco, Richard. (1990). *A Path Where No Man Thought: Nuclear Winter and the End of the Arms Race*. New York: Random House.

Shklovskii, I. S., & Sagan, Carl. (1966). *Intelligent Life in the Universe* (Paula Fern, Trans.). San Francisco: Holden-Day.

Soviet-American Conference on the Problems of Communication with Extraterrestrial Intelligence. (1971). *Communication with Extraterrestrial Intelligence*, edited by Carl Sagan. Byurakan Astrophysical Observatory.

U.S. National Research Council Ad Hoc Panel on Planetary Atmospheres. (1961). *The Atmospheres of Mars and Venus*. Report prepared by William W. Kellogg & and Carl Sagan. Washington, DC: National Academy of Sciences-National Research Council.

OTHER SOURCES

A&E Biography: Carl Sagan. (2000). Produced by Brooke Runnelle. [VHS video]. Arts & Entertainment.

Broad, William J. (1998, November 30). Even in Death, Carl Sagan's Influence Is Still Cosmic. *Science: The New York Times*. Retrieved January 27, 2004, from http://www.nytimes.com/library/national/science/120198sci-sagan.html.

Chang, Kenneth. (2004, January 25). Images from 2nd Mars Rover Show "Bizarre Alien Landscape." *The New York Times*. Retrieved January 25, 2004, from http://www.nytimes.com/2004/01/25/science/space/25MARS.html.

Connell, Kathleen. (2002, July 12). Interview with Ann Druyan and Steven Soter: Sagan as Teacher, as Collaborator. *Astrobiology Magazine*. Retrieved October 8, 2002, from http://www.astrobio.net/news/modules.php?op=modload&name=Sections&file=index&req=viewarticle&artid=30&page=4.

Davidson, Keay. (1999). *Carl Sagan: A Life*. New York: J. Wiley.

Dawkins, Richard, et al. (1997). The Darkened Cosmos: A Tribute to Carl Sagan. *Skeptical Inquirer, 21*(2), 5–15.

Drake, Frank, & Sobel, Dava. (1992). *Is Anyone Out There? The Scientific Search for Extraterrestrial Intelligence*. New York: Delacorte Press.

Druyan, Ann. (2000, November). A New Sense of the Sacred: Carl Sagan's "Cosmic Connection." *The Humanist*. Retrieved June 26, 2003, from http://www.findarticles.com/cf_dls/m1374/6_60/78889720/p1/article.jhtml.

Druyan, Ann. (2003, November/December). Ann Druyan Talks about Science, Religion, Wonder, Awe . . . and Carl Sagan. *Skeptical Inquirer: The Magazine for Science and Reason, 27*(6), 25–30.

Druyan, Ann. (2004, May 7). Conversation with Kit Moser.

Ehrlich, Robert. (2003). *Eight Preposterous Propositions: From the Genetics of Homosexuality to the Benefits of Global Warming*. Princeton, NJ: Princeton University Press.

Ezell, Edward Clinton, & Ezell, Linda Neuman. (1984). *On Mars: Exploration of the Red Planet*. NASA SP-4212. Washington, DC: NASA History Office.

Gardner, Martin. (1957). *Fads and Fallacies in the Name of Science*. New York: New American Library.

Ginenthal, Charles. (1995). *Carl Sagan & Immanuel Velikovsky*. Tempe, AZ: New Falcon Publications.

Henahan, Sean. (1996). From Primordial Soup to the Prebiotic Beach: An Interview with Exobiology Pioneer, Dr. Stanley L. Miller, University of California San Diego. *Access Excellence*. Retrieved December 26, 2003, from http://www.accessexcellence.org/WN/NM/miller.html.

Horowitz, Paul. (1996). Kidnapped by UFOs? Interview with Paul Horowitz, Physicist, Harvard University. *NOVA*. Retrieved March 10, 2004, from http://www.pbs.org/wgbh/nova/aliens/paulhorowitz.html.

Klass, Philip J. (1974). *UFOs Explained*. New York: Random House.

Klass, Philip J. (1983). *UFOs: The Public Deceived*. Buffalo, NY: Prometheus Books.

Mackay, Charles. (1993). *Extraordinary Popular Delusions and the Madness of Crowds*. Original publication, 1841. (Reprint.). New York: Barnes and Noble.

Margulis, Lynn. (1999). Participant, *Cajál Conference on Consciousness*, Zaragoza, Spain. Retrieved December 30, 2003, from http://cajal.unizar.es/eng/part/Margulis.html.

NOVA. (1996). Interview with Carl Sagan: Author, Astronomer. Retrieved June 23, 2001, from http://www.pbs.org/wgbh/nova/aliens/carlsagan.html.

Obst, Lynda. (1996, February). Valentine to Science. Interview with Carl Sagan. *Interview*. Retrieved June 26, 2003, from http://www.findarticles.com/cf_dls/m1285/n2_v26/18082728/p1/article.jhtml.

Ponnamperuma, Cyril, (Ed.). (1972). *Exobiology*. Amsterdam: North-Holland.

Poundstone, William. (1999). *Carl Sagan: A Life in the Cosmos*. New York: Henry Holt.

Sheaffer, Robert. (1981). *The UFO Verdict: Examining the Evidence*. Buffalo, NY: Prometheus Books.

Sheaffer, Robert. (1998). *UFO Sightings: The Evidence*. Amherst, NY: Prometheus Books.

Terzian, Yervant, & Bilson, Elizabeth, (Eds.). (1997). *Carl Sagan's Universe*. New York: Cambridge University Press.

Tyson, Neil deGrasse. (1997, February 27). Reminisces of Carl Sagan: A Celebration of Carl Sagan's Life. Retrieved February 13, 2003, from http://research.amnh.org/users/tyson/speeches/CarlSaganEulogy.html.

Weed, William Speed, & Ferris, Timothy. (2000, March 21). *Salon Brilliant Careers*. Retrieved March 8, 2004, from http://archive.salon.com/people/bc/2000/03/21/ferris/.

INDEX

About the Authors

RAY SPANGENBURG, with Kit Moser has written many books on science and the history of science, including *Niels Boher: Gentle Genius of Denmark* (1995) and the "On the Shoulders of Giants" series (1994).

KIT MOSER, with Ray Spangenburg, has written many books on science and the history of science, including *Niels Boher: Gentle Genius of Denmark* (1995) and the "On the Shoulders of Giants" series (1994).